A GENETIC SWITCH

THIRD EDITION
Phage Lambda Revisited

Mark Ptashne
Memorial Sloan-Kettering Cancer Center
New York

COLD SPRING HARBOR LABORATORY PRESS
Cold Spring Harbor, New York • http://www.cshlpress.com

A GENETIC SWITCH
Phage Lambda Revisited
Third Edition

Publisher	John Inglis
Acquisition Editors	Alexander Gann and John Inglis
Developmental Editor	Alexander Gann
Project Coordinator	Maryliz Dickerson
Production Editor	Melissa Frey
Desktop Editor	Susan Schaefer
Production Manager and Interior Designer	Denise Weiss
Cover Designer	Mike Albano

Library of Congress Cataloging-in-Publication Data

Ptashne, Mark.
  A genetic switch : phage lambda revisited / by Mark Ptashne.-- 3rd ed.
      p. ; cm.
  Includes bibliographical references and index.
  ISBN 0-87969-716-4 (alk. paper)
  1. Bacteriophage lambda. 2. Genetic regulation. 3. Viral genetics.
    [DNLM: 1. Bacteriophage lambda--genetics. 2. Gene Expression Regulation. 3. Genes, Viral. 4.
  Lysogeny. 5. Repressor Proteins. QW 161.5.C6 P975g 2004] I. Title.
    QR342.P83 2004
    579.2'6--dc22

                            2004000803

10 9 8 7 6 5 4 3

*This book is dedicated to my parents
and to Matt Meselson*

CONTENTS

PREFACE TO THE THIRD EDITION

B iologists work on systems that have evolved. This gives us hope that any given case can be understood reductively. Nature built the system in steps, each step making an improvement on the previous version and so, this line of thought goes, the investigator can take it apart, study it in bits, and, perhaps, see how it all fits together. These notions have proved apt in studying gene regulation in the bacterial virus lambda.

The first edition of this text (published in 1986) described a series of protein-protein and protein-DNA interactions that effect two alternative patterns of gene expression in phage lambda. This "genetic switch" (as we called it) ensures an efficient change from one pattern to the other in response to an environmental signal.

This new edition is prompted by discoveries (none due to my efforts) that add to, rather than reformulate, the earlier story. The most striking of these describes interactions between proteins binding to widely separated sites on DNA, and shows how these interactions make the switch even more efficient than we had thought. Others probe more deeply than before, showing us, in clear molecular terms, how an activator of transcription works. Other experiments begin to dissect how the complex lambda gene regulatory network might have evolved, and still another reveals an enzymatic function carried by the repressor itself.

These new developments are described here in a chapter (Chapter 5) added to a reprint, only slightly modified, of the first edition. Those already familiar with the original description of the switch can learn about these developments by going directly to the new chapter; others can learn the whole story by reading the complete book as it now stands. The extended discussion of gene regulation in eukaryotes, added as part of the second edition but omitted here, has been reformulated and placed in a larger context in the book *Genes and Signals* (Ptashne and Gann 2002 Cold Spring Harbor Laboratory Press). The preface to the second edition has also been omitted here.

The new chapter illustrates, as did the original text, that uncovering interactions between components of the switch, and assessing their biological significance, required combining genetic with biochemical studies, and in some cases with biophysical and structural studies as well. Given how weak some of the inter-actions are—for example, those mediating cooperative binding of proteins to DNA—it is hard to see how any other approach would have worked. Our picture of the switch has been formulated by studying the component reactions in isola-tion and then piecing them together. The fact that a step-wise approach has pro-duced a picture of such (seeming) completeness and coherence is in itself (it seems to me) remarkable.

Strictly speaking, this book was, and remains, mistitled—"An Epigenetic Switch" would be more appropriate. An epigenetic change in the state of gene expression is one that persists in the absence of the original signal (or event), and that involves no change in DNA sequence. The maintenance of lysogeny over many cell divisions following its establishment (see Chapter 3) is a classic example of epigenesis.

There are important matters of gene regulation in lambda, currently under study, that I do not discuss in the new chapter. The mechanisms of action of the "anti-terminators" N and Q (introduced in Chapter 3) are two examples. Systems biologists have applied quantitative modeling approaches to this or that aspect of the lambda lifestyles and the transition between them, but I have not attempted to review these matters either. References to some of these studies are to be found at the end of Chapter 5.

All the figures from this book (along with those from *Genes and Signals*) are available on the Web Site www.genesandsignals.com.

Once again generous people have come to my rescue, offering facts, figures, advice, and admonishments. I want to thank particularly Alex Gann, whose encouragement and advice at every step was crucial, as well as Seth Darst, Ian Dodd, Ann Hochschild, Deepti Jain, Leemor Joshua-Tor, Mitchell Lewis, John Lit-tle, and Keith Shearwin, each of whom contributed figures or unpublished results, commented extensively on the text, or both. I also thank Alan Campbell, Simon Dove, Richard Ebright, Barry Egan, Drew Endy, Lenny Guarente, Barry Honig, Sandy Johnson, Tom Laue, Wendell Lim, Richard Losick, David Senear, Richard Treisman, and Jose Vilar, each of whom also helped me think certain things through. Mary Jo Wright's typing and organizing saved the day on more than one occasion. Hans Neuhart accurately redrew the old figures as well as the new ones, lickety-split, while the efforts of Denise Weiss, Melissa Frey, and Susan Schaefer at CSHL Press were invaluable in putting the book together.

MARK PTASHNE
New York, NY

January 2004

PREFACE TO THE
FIRST EDITION

This book is about one of nature's simplest organisms—a virus that grows on a bacterium. It describes the results of some 25 years of research on the question of how the virus called lambda (λ) uses its genes—its DNA—to direct its growth.

Why has so much effort been expended on studying one virus? This is a fair question ... after all, every case in biology is at least partly accidental and special. The workings of every organism have been determined by its evolutionary history, and the precise description we give of a process in one organism will probably not apply in detail to another. The answer is to be found in the context of the fundamental biological process called development.

Briefly put, the issue is as follows: all cells of a given individual organism inherit the same set of blueprints in the form of DNA molecules. But as a higher organism develops from a fertilized egg a striking variety of different cell types emerges. Underlying the process of development is the selective use of genes, the phenomenon we call gene regulation. At various stages, depending in part on environmental signals, cells choose to use one or another set of genes, and thereby to proceed along one or another developmental pathway. What molecular mechanisms determine these choices?

The λ life cycle is a paradigm for this problem: the virus chooses one or another mode of growth depending upon extracellular signals, and we understand in considerable detail the molecular interactions that mediate these processes. We believe that analogous interactions are likely to underlie many developmental processes; by establishing a description for the particular case of λ, we develop ideas that inform other studies even though no other case looks exactly like λ.

The Introduction to this book describes some basic facts about genes and how they work. It is designed to enable the reader with a modest knowledge of molecular biology to understand the first three chapters.

These chapters then describe the process of λ's development from three perspectives: from a distance, showing the overall pattern of events involved in the interaction between the virus and its host bacterium; more closely, describing in coarse molecular terms a key event in the process; and very closely, showing precise molecular interactions. The basic concepts at each level are presented in these chapters in a series of pictures without reference to experiments. At various points, our understanding of λ is related to developmental processes and gene control in other organisms.

Chapter 4, more technical than the first three chapters, describes the principles of some of the key experiments. The experiments and the arguments based on them are easier to follow if one knows the answers as outlined in the first three chapters.

The reader will see that we now have a coherent understanding of many aspects of gene regulation in λ. Our integrated picture accounts for the experimental observations, and more importantly, predicts results of new experiments. This degree of rigor is achieved, in part, because very few of our models depend upon any isolated experimental observation; rather, they are based on integrated sets of experiments carried out both in the test tube and in living cells.

The book taken as a whole is thus a case study that shows how biochemical and genetic experiments construct a view of part of the world. I have avoided an historical approach—a different and longer exposition would be required to describe how our understanding developed.

One of the charms of molecular biology is that the answers it provides to fundamental questions for the most part can be easily visualized. Simple pictures will usually do, and only rarely need we invoke abstruse ideas. Our goal is to understand gene regulation in terms of the interaction of molecules. A glimpse of the characteristic sizes and shapes of these molecules often reveals, or helps us remember, how they work. Please take the pictures in this book seriously, but for what they are—a summary of our current views. I fully expect that in the years to come they will be redrawn as our understanding deepens.

For their friendly advice on this project I want to thank, in addition to students and colleagues in my laboratory and at Harvard University, Alison Cowie, Nick Cozzarelli, Norm Davidson, Hatch Echols, Gary Gussin, Gerhard Hochschild, Will McClure, Russ Maurer, Howard Nash, Jeff Roberts, Bob Schleif, Hamm Smith, John Staples, Jim Watson, Adam Wilkins, Keith Yamamoto, Michael Yarmolinsky, Patricia Zander, and especially Sandy Johnson. Bernard Hirt suggested the idea in the first place.

Three people played especially important roles in producing this book—Ben Lewin coached and edited with style and understanding; Carol Morita created the illustrations imaginatively and quickly; and Carol Nippert typed and retyped, organized and reorganized, superbly.

MARK PTASHNE
Cambridge, MA

January 1986

INTRODUCTION

ome 40 years ago, André Lwoff and his colleagues at the Institut Pasteur in
Paris described a dramatic property of a certain strain of the common intesti-
nal bacterium *Escherichia coli*. If irradiated with a moderate dose of ultraviolet light these bacteria stop growing, and some 90 minutes later they lyse (burst),
spewing a crop of viruses called λ into the culture medium. The viruses are also
called bacteriophages—bacteria eaters—or simply phages. The liberated λ phages
multiply by infecting fresh bacteria. Many infected bacteria soon lyse and produce
new phage but some survive and carry λ in a dormant form. These bacteria grow
and divide normally until the culture is once again irradiated—then each of these
progeny bacteria, like those with which we started, lyses and yields a new crop of
λ phages. Figure I.1 shows a picture of the virus and its host.

Lwoff and his colleagues François Jacob and Jacques Monod realized that this
switching between two states of the virus—from the dormant form in the dividing
bacterium to the activated form in the irradiated bacterium—is a simple example
of a fundamental biological process: the turning on and off of genes.

Genes determine the structures of molecules that constitute living cells. At any
given time, each cell—be it bacterial or human—uses only a subset of its genes to
direct production of other molecules, and we say that these expressed genes are
on and those not expressed are *off*. We say in other words that the expression of
these genes is regulated.

As an example, consider the development of a person from a fertilized egg. As
this cell and then its descendants divide—a process repeated millions of times—
each new cell receives an identical set of genes. Nevertheless some cells (for example, hair cells) look and act different from others (for example, skin cells) because
different genes are turned on in the different cells. The essential point is that
although genes are transmitted unchanged (ignoring a few exceptions) from
parental to progeny cells, they may be expressed differently in various cells.

Gene expression is regulated not only during development but also during the
lifetime of the differentiated cell. For example, a skin cell changes color when
exposed to sunlight. The structure of the pigmentation gene does not change in
response to the light; rather the extracellular signal, the light, turns the gene on. To

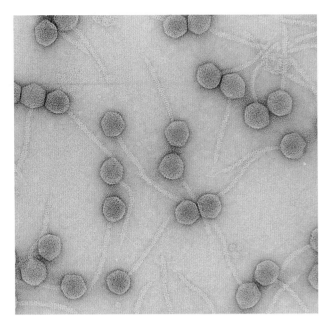

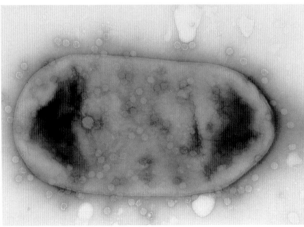

Figure I.1. Electron micrograph of λ and of an *E. coli* bacterium being infected with λ. The diameter of the lambda head is ~600 Å; magnifications are ×100,000 (upper) and ×8000 (lower). Photographs kindly provided by Roger Hendrix.

take another example of gene regulation: cancer cells multiply under conditions that their normal counterparts do not, partly because certain genes are on (or off) when they should not be.

Biologists have long wished to know how genes are turned on and off normally during development and, aberrantly, in diseased states. We are interested both in the molecular mechanisms of gene regulation and in how these mechanisms are integrated into interlocking circuits that ensure the orderly turning on and off of sets of genes. We wish to understand which steps are controlled by internal cellular programs and which by extracellular signals.

Returning to our original example—phage λ—we can now appreciate the insight of those earlier workers who recognized in the growth of this virus—in particular its ability to grow in two different modes—a revealing example of the regulation of gene expression. Bacteria and their phages multiply quickly, and it is possible to combine genetics with biochemistry to analyze gene regulation much more efficiently than is possible with cells of higher organisms. We turn now to a brief description of genes and how they work.

Our starting point is the structure of a gene, a piece of DNA, by itself an inert molecule. But DNA carries information in the sequence of bases along its two strands. The four bases—adenine (A), thymine (T), guanine (G), cytosine (C)—are attached to the two intertwined backbones so that A on one strand is always paired with T on the other, and similarly a G is always paired with a C.

Figure I.2 shows three representations of a segment of DNA. In the first the sequence of 24 base pairs is written on two lines, the bases along the top line representing one strand of DNA—the "top" strand—and those along the bottom line the other strand. We say the sequences of the bases along the strands are complementary because the base pairing rules require that the sequence of either strand dictates that of the other.

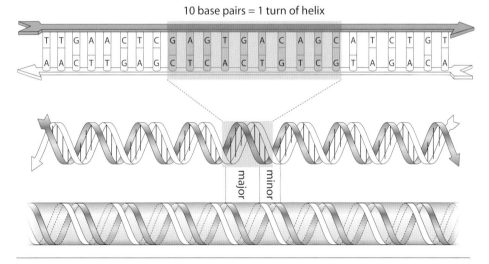

Figure I.2. Different ways to visualize DNA. The representation on the top emphasizes the complementary base pairing. The second shows the shape of the double helix with its major and minor grooves. The polarity of the strands is indicated by the arrow heads and tails. The two grooves are also seen in the "barber pole" representation on the third line. (The precise number of base pairs per turn of the helix is 10.5 and not 10.)

In the second representation in the figure, some 80 base pairs are modeled in the typical double helical form. Note that because one turn of the helix comprises about 10 base pairs, these 80 base pairs form about eight helical turns. Two extended grooves of different widths—the major and the minor grooves—wind around the surface of the helix. This representation shows that the backbones run along the outside of the helix while bases face inwards.

Finally, in our third representation, a DNA molecule looks like a barber pole bearing traces of the two strands of the helix. This form, in which the various structural features of the helix are simplified but kept in proportion, is used throughout Chapter One.

The arrow heads and tails on each strand in the figure indicate another feature of the DNA double helix: each strand has a polarity or directionality, and the two strands run in opposite directions. As Figure I.3 shows, this strand polarity is a consequence of the fact that the chemical linkages along the backbone of each strand are asymmetric. For reasons explained in the figure, the arrow heads and tails are called, respectively, the 3′ and 5′ ends of the chain.

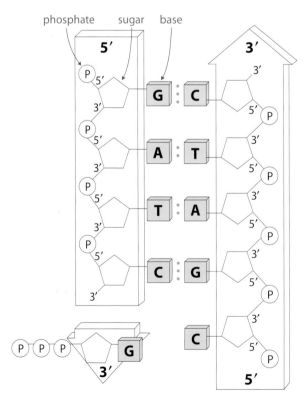

Figure I.3. A closer look at DNA. The backbone of each DNA strand comprises alternating sugars and phosphates. Directionality is defined by the way each sugar is attached to one phosphate by a 5′ linkage, and to another phosphate by a 3′ linkage. Each sugar is also attached to a base that is paired with its complementary partner on the other strand. Three interactions—hydrogen bonds—form between G:C base pairs, and two between A:T pairs. A guanine base, attached to a sugar-phosphate, will add to the 3′ end of one of the strands, and two phosphate groups will be removed in the process.

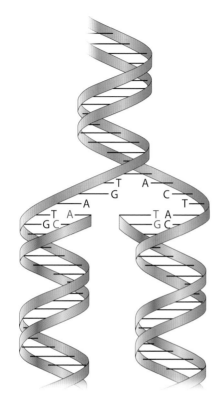

Figure I.4. DNA replication. In the stem of the wishbone the DNA strands have not yet come apart. In the arms the strands are separated and are used as templates for formation of the complementary strands. As the strands are copied, A pairs with T, and G pairs with C.

The principle of sequence complementarity explains how genes are faithfully duplicated and, as we shall see in a moment, how they are expressed as well. Figure I.4 shows that as DNA replicates its two strands unwind, and each is sequentially copied using the rules of base pairing. In this process each parental strand is used as a template for formation of its complement, and every time a cell divides its two daughters receive replicas of the parental genes.

A typical gene consists of a sequence of about 1000 base pairs (100 helical turns) within a much larger DNA molecule. The first step in gene expression is always the same: the sequence along one of its strands is copied, or transcribed, into a linear molecule called RNA. As indicated in Figure I.5, the sequence of bases along the RNA is identical with that along one of the DNA strands and the complement of that along the other (the template). For our purposes it suffices to define a gene as on if it is being copied into RNA, and off if not.

Some RNA molecules are end products that function directly in the cell. Others—called messenger RNAs (mRNAs)—specify the designs of molecules called proteins. There are many kinds of proteins, including, for example, structural com-

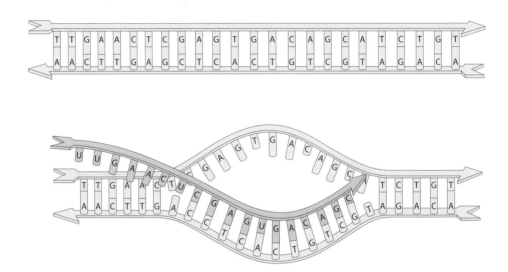

Figure I.5. Complementarity of mRNA with DNA. In RNA the base U is the equivalent of the base T in DNA. Thus the sequence of this mRNA sequence is complementary to the DNA sequence of the bottom strand. This mRNA, as well as those of the next two figures, is growing in the rightward direction.

ponents, antibodies, and enzymes. The latter carry out the actual work of the cell, including transcribing and duplicating DNA.

A protein consists of a string of units called amino acids, whose sequence is determined by the sequence of bases along the gene. The mRNA is read (translated) sequentially, beginning at its 5´ end. Each successive group of three bases specifies one of 20 amino acids to be added to the growing protein chain. The first amino acid of the protein chain is called the protein's amino terminus, the last its carboxyl terminus.

Each protein folds into a characteristic shape determined by its amino acid sequence, typically forming an irregular globule whose surface is marked by cavities and protuberances. Here and in Chapter One we will discuss examples of proteins pictured as blobs, and in Chapter Two we will see how characteristic surface features of a group of proteins involved in gene regulation determine their function.

Genes are transcribed into mRNA by an enzyme called RNA polymerase. The process begins with the binding of this enzyme near the beginning of a gene to a site called a promoter, a region extending over some 60 base pairs. Figure I.6 shows an RNA polymerase molecule about to bind to a promoter of a typical bacterial gene. We do not know much about the shape of this enzyme but we know its approximate size.

Following the initial binding, polymerase travels away from the promoter along the gene, synthesizing the mRNA as it moves, as in Figure I.7. Note that as polymerase moves it continuously unwinds and then rewinds successive short regions of

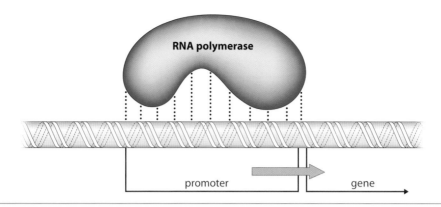

Figure I.6. The promoter and RNA polymerase. RNA polymerase, when bound to the promoter at the beginning of the gene, covers six turns of the helix, about 60 base pairs. This simplified representation of polymerase ignores the fact that this large enzyme consists of several chains of different amino acid sequence. The arrow shows the direction of transcription, which begins at the base pair just above the arrow's tail.

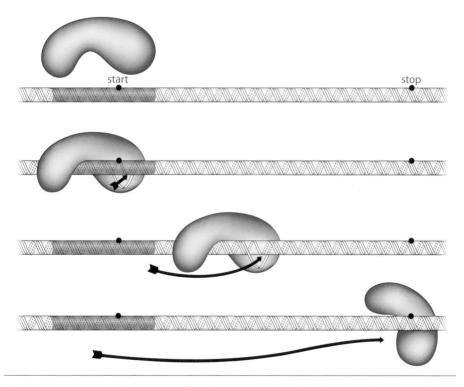

Figure I.7. Transcription of a gene. As the polymerase moves along the DNA, the unwound DNA segment, about one helical turn in length, is always in contact with the same part of the polymerase. At any given instant the template base being copied lies in the unwound region and is about 20 base pairs from polymerase's nose. When polymerase reaches the stop signal it falls off the DNA and releases the mRNA.

7

DNA. This transient unwinding separates the base pairs so that the sequence along one of the strands becomes a template for formation of the complementary mRNA.

Each promoter points polymerase in either one or the other direction along the DNA helix, and as the enzyme moves in a given direction it copies only one of the strands into mRNA. The polarity of the mRNA chain is opposite that of the DNA template strand.

As represented in Figure I.8, two polymerase molecules moving in opposite directions copy different strands. The polymerase moving leftward copies the "top" strand as defined in the representation of DNA at the top of Figure I.2. When we consider the virus λ in greater detail we shall see that its single long DNA molecule consists of many genes, some of which are transcribed rightward, some leftward.

RNA polymerase can be helped or hindered in its attempt to transcribe a gene by "regulatory" proteins that bind to sites on the DNA called operators. A negative regulator prevents transcription, and a positive regulator increases (stimulates) transcription of a gene. In Chapter One we will explain the workings of one regulatory protein—the λ repressor—that is both a positive and negative regulator of transcription.

We will often say that a particular regulatory protein binds to a specific operator site (or sites) on a DNA molecule. We mean by this that the protein is usually to be found there, but that it can quite readily (and often does) fall off that site. Whether another identical protein quickly binds again depends upon its concentration and its affinity for the DNA site.

A DNA molecule may have more than one site able to bind a particular protein. Such sites can vary in the strength with which they bind the protein. If one site is weaker than another then, at any given instant, and at low protein concentrations, the stronger site is more apt to have a protein bound to it. But, as illustrated in Figure I.9, at high protein concentrations the difference in affinities would be ignored and both sites would usually be filled. This state of dynamic equilibrium obtains because the bonds involved in protein-DNA interactions are much weaker than those (for example) that hold the links of the protein chain together.

The first four chapters of this book are designed as follows. Chapter One describes how the regulatory proteins λ repressor and Cro bind to DNA and interact with RNA polymerase in a manner that determines which promoters will be

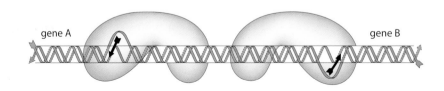

Figure I.8. Divergent transcription. The sequence of the mRNA of gene B is complementary to one of the DNA strands, and that of gene A is complementary to the other DNA strand.

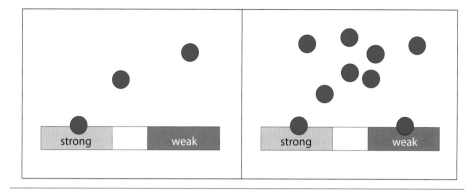

Figure I.9. Binding of a regulatory protein to a strong and a weak operator site. A regulatory protein fills only the strong operator site at low concentration, but fills both operator sites at high concentration. Put another way, the protein first binds the strong site, then the weak site, as its concentration increases.

used to initiate transcription. These components, constituting a "genetic switch," control the remarkably efficient change in gene expression that is triggered when a bacterium bearing a dormant λ is irradiated with ultraviolet light.

To understand the logic of the switch mechanism at this stage, we need to invoke knowledge of molecular structure, but only crudely. We picture DNA as a rigid rod bearing sites to which RNA polymerase and the regulatory proteins, pictured as spheres or dumbbells, bind to promoters and operators, turning genes on and off. Many simple pictures are used to portray each component and each interaction separately so that the mechanisms become transparent.

Chapter Two describes the structures of regulatory proteins in much greater detail than Chapter One. It elucidates a simple and general mechanism, with some variations, by which these proteins recognize specific base sequences among millions of base pairs. The structure of a regulatory protein is shown to be complementary to part of the DNA structure; if the sequences are correct, the two molecules fit together like lock and key. Analysis at this level of detail enables us to surmise how a regulatory protein can turn a gene on or off.

Chapter Three traces the patterns of λ gene activity as the phage lyses the cell, or alternatively, as it becomes dormant in the cell. The first few steps of gene regulation that occur upon λ infection are identical for both pathways. At the critical step the state of the host is "sensed" by a phage regulatory protein that determines which pathway subsequent events shall take. This "decision" is an instructive example of how the environment can influence gene regulation during development. Once initiated, the regulatory sequence along each pathway is a cascade—groups of genes are turned on and off sequentially according to an internally determined program.

Chapter Four outlines some of the experimental bases for many of the pictures in Chapter One and Chapter Two. I have simplified some of the arguments and have not tried to be complete. Nevertheless the reader unfamiliar with the techniques used in experimental molecular biology will find this chapter more difficult to follow. This chapter need not be read beginning to end, but rather might be dipped into according to the reader's special interest. The chapter concludes by noting some unsolved problems.

The main purpose of this book is to provide an account of the mechanisms used to regulate individual λ genes and of how these mechanisms interconnect to form regulatory networks. From a wider perspective, we believe that a central aspect of the general question of development—how a complex organism develops from a fertilized egg—involves elaborate networks of differentially regulated genes. At various points, particularly at the end of Chapter Three, I have therefore drawn some parallels between developmental processes and gene regulation in λ and in higher organisms.

A note on nomenclature: genes are denoted with italicized letters, usually but not always in lowercase, for example, *cro, recA, lexA, N, Q*; and their protein products with Roman letters, first letter capitalized, for example, Cro, RecA, LexA, N, Q. Sometimes for historical reasons, the protein has a special name, for example, repressor, encoded by the gene *cI*.

CHAPTER ONE

THE MASTER ELEMENTS OF CONTROL

The genes of phage λ constitute a single DNA molecule—its chromosome—wrapped in a protein coat (Figure 1.1). The coat is an elaborate structure with a head and a tail, together composed of some 15 different proteins, all encoded by the λ chromosome. The phage particle is infectious: it attaches by its tail to the surface of an *E. coli* cell, drills a hole in the cell wall, and squirts its chromosome into the bacterium, leaving its coat behind. Lambda is an obligate parasite—it must inject its DNA into the bacterium to multiply.

One of two fates awaits the λ-infected bacterium, as illustrated in Figure 1.2. In some cells the phage chromosome enters the lytic cycle: various sets of phage genes turn on and off according to a precisely regulated program, the λ chromosome is extensively replicated, new head and tail proteins are synthesized, new phage particles are formed within the bacterium, and some 45 minutes following infection the bacterium lyses and releases about 100 progeny phage.

In other cells the injected phage chromosome lysogenizes its host: all but one of the phage genes are turned off, and one phage chromosome—now called prophage—becomes part of the host chromosome. As the lysogen—the bacterium bearing the prophage—grows and divides, the prophage is passively replicated and quiescently distributed to the progeny bacteria, all the while as part of the host chromosome. This process may be repeated indefinitely, and if unperturbed these growing and dividing lysogenic bacteria only very rarely produce phage.

When irradiated with ultraviolet light, however, virtually every lysogen in the population will lyse and produce a new crop of λ. The ultraviolet light turns on, or induces, previously inert phage genes and lytic growth ensues. Many agents, like ultraviolet light, induce lytic growth in lysogens by damaging the host DNA. The phage chromosome uses bacterial enzymes to "sense" the impending demise of its host—an effect of the inducing agent—and, abandoning its previously useful scheme of passive replication, it enters the lytic pathway.

This chapter explains how the phage genes are maintained stably in the lysogenic state and are then switched with high efficiency to a second state—lytic growth—upon exposure of a lysogen to an inducing signal. Viewed broadly, the switch works as follows.

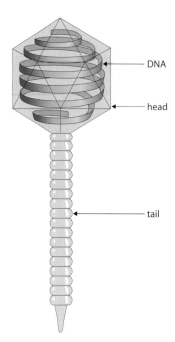

Figure 1.1. A λ particle. The λ chromosome—some 50,000 base pairs of DNA—is wrapped around a protein core in the head.

In a lysogen the single phage gene that is expressed encodes the λ repressor. This protein is both a positive and a negative regulator of gene expression. By binding to just two operators on λ DNA it turns off all the other phage genes as it turns on its own gene. (How the repressor-encoding gene gets turned on in the first place—immediately following infection when there is no repressor present—will be explained in Chapter Three.)

Although there is only one prophage in a lysogen there are about 100 active molecules of repressor, and the excess repressor is free to bind to any additional λ chromosomes that might be injected into the cell. This has the result illustrated in Figure 1.3: λ cannot grow lytically on a λ-lysogen. The lysogen is said to be immune to λ infection.

Ultraviolet irradiation of lysogens inactivates repressor. As a result a second phage regulatory protein—Cro—is synthesized. Cro, which promotes and is required for lytic growth, also binds DNA—in fact it binds to the same operator sites as does repressor, but with opposite physiological effects. These two regulatory proteins, together with RNA polymerase and their promoter and operator sites on DNA, constitute the switch.

Simplifying somewhat we say that the switch has two positions: in the first (that is, in a lysogen) the repressor gene is on, but the gene encoding Cro is off; and in the second (that is, during lytic growth) the *cro* gene is on but the repressor gene is off. We now describe the approximate sizes and shapes of the switch components and how some of them interact.

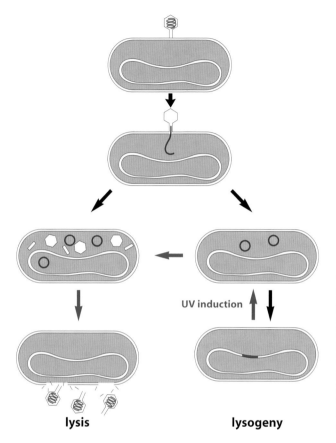

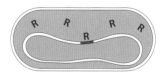

lysis   lysogeny

Figure 1.2. Growth of phage λ. The injected λ chromosome may either lyse or lysogenize the host. Ultraviolet irradiation of a lysogen induces lytic growth. Induction of lysogens was first demonstrated for a prophage of the bacterium *Bacillus megaterium.*

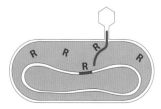

Figure 1.3. Immunity of a λ-lysogen. Lambda phages inject their chromosomes into a λ-lysogen, but repressor molecules (R) immediately turn off the genes of these "superinfecting" chromosomes, just as they turn off the genes of the prophage. Immunity is thus caused by the same repressor that maintains the prophage in its dormant state.

COMPONENTS OF THE SWITCH

DNA

The genes that encode repressor and Cro—namely, *cl* and *cro*—are adjacent on the λ chromosome. These genes are transcribed in opposite directions, divergently, as shown in Figure 1.4. The start points of transcription of these two genes are separated by 80 base pairs, and in this region lie two kinds of sites—promoter and operator—to which protein components of the switch can bind.

Figure 1.4 shows that each of our genes *cl* and *cro* has its own promoter. The *cl* promoter, called P_{RM}, points polymerase leftward, and the *cro* promoter, P_R, points polymerase rightward. The surfaces of the DNA molecule that comprise these two promoters are shaded in the figure. Note that these two promoters are adjacent but do not overlap.

Three adjacent sites—O_R1, O_R2, and O_R3—comprise O_R, the right operator of λ, also shown in Figure 1.4. Repressor and Cro bind to these sites to regulate the activities of the two promoters. Note the order of the operator sites as shown in the figure; each site overlaps one or the other promoter, or, in the case of O_R2, both of the promoters.

Each of the three individual operator sites is 17 base pairs; their sequences, although similar, are not identical, and the regulatory proteins can distinguish between them. For example, considering any two operator sites, one might have a higher affinity than the other for Cro. This would mean that, at some specified concentration of the protein, repeated snapshots of the operator would reveal that a Cro molecule was

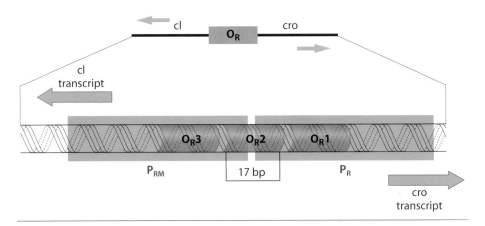

Figure 1.4. A short segment of the λ DNA molecule. Two back-to-back promoters (P_{RM} and P_R) send polymerase traveling in opposite directions—leftward to transcribe the repressor gene (*cl*) and rightward to transcribe the *cro* gene. The tripartite right operator (O_R) overlaps the two promoters. Each of the three parts of the operator is called an operator site.

found more frequently at the stronger binding site. At higher protein concentrations, the snapshots would reveal Cro usually bound to the second site also.

A note on nomenclature: the rationale behind the nomenclature of the switch elements will not become entirely clear until later, particularly in Chapter Three. Briefly, P_R and O_R stand for the *right* promoter and operator. There are also a left promoter and operator *(P_L and O_L)* the properties and roles of which are considered in later chapters. The name *cl* distinguishes this gene from genes *cll* and *clll; cro* stands for *c*ontrol of *r*epressor and *o*ther genes, because that is what Cro does. P_{RM} stands for *p*romoter of *r*epressor *m*aintenance to distinguish it from a related promoter that we discuss in Chapter Three.

RNA Polymerase

RNA polymerase, the enzyme that transcribes genes from DNA to RNA, is provided by the bacterial host. When bound at P_R, polymerase is poised to transcribe (rightward) the *cro* gene, and when at P_{RM}, polymerase can move leftward, to transcribe the *cl* gene.

Figure 1.5 illustrates the positions that would be occupied *if* both promoters were occupied by polymerases at the same time. In fact, these two promoters are never occupied simultaneously in the cell—depending on the position of the switch, polymerase can bind to one or to the other. Thus, as the figure suggests, in the presence of repressor polymerase might bind to P_{RM} but never to P_R, and vice versa in the presence of Cro. One of the important questions of this chapter is— how do repressor and Cro have these opposite effects?

An important difference between the promoters P_R and P_{RM} is that RNA polymerase binds and begins transcription at the former without the aid of any positive regulatory protein. In contrast, polymerase works efficiently at P_{RM} only if helped by an activator protein, a role played by the repressor.

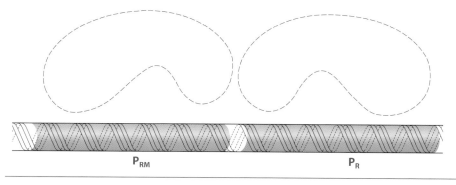

Figure 1.5. An event that never occurs. RNA polymerase molecules could occupy both P_R and P_{RM} on the same DNA molecule. But this does not happen—in the presence of repressor, polymerase may occupy P_{RM}, but never P_R, and in the presence of Cro, P_R may be occupied, but not P_{RM}.

The Repressor

The repressor—shown in Figure 1.6—is a protein of 236 amino acids that folds into two nearly equal-sized blobs, called domains, connected by a string of 40 amino acids. The domains are called amino and carboxyl—the former includes the first amino acid of the chain and the latter the last amino acid of the chain.

    Two of the chains of Figure 1.6, called monomers, associate to form dimers as shown in Figure 1.7. The dimer forms largely because of contacts between carboxyl domains, the amino domains contributing only slightly to this reaction. In a lysogenic cell about 95% of the repressor molecules are associated as dimers as indicated in Figure 1.8.

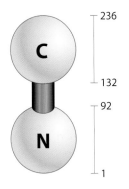

Figure 1.6. The λ repressor. The amino domain of repressor comprises amino acids 1–92 and its carboxyl domain of residues 132–236. The remaining 40 amino acids connect the two domains.

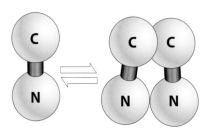

Figure 1.7. Dimerization of λ repressor. Repressor monomers associate to form dimers which, in turn, dissociate to monomers. We say that repressor monomers are in equilibrium with dimers. As the concentration increases, a larger fraction of the repressor is present as dimer.

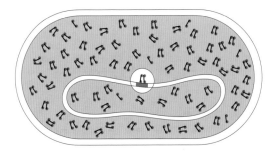

Figure 1.8. Repressor in a lysogen. Most of the repressor in the lysogenic cell is in the dimer form. The single long *E. coli* chromosome contains one integrated prophage to which repressors are bound tightly. The rest of the dimers associate loosely with other parts of the bacterial chromosome or float freely in the cell.

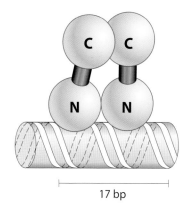

Figure 1.9. A repressor dimer bound to one 17 base pair operator site. Each amino domain is centered on a segment of the major groove, a point we return to in Chapter Two.

17 bp

Repressor dimers use their amino domains to bind to DNA as shown in Figure 1.9, and each of the three sites in O_R can bind one repressor dimer. The repressor dimer binds along one face of the helix at each site.

Cro

Cro—shown in Figure 1.10—is made of only 66 amino acids, folded into a single domain. The affinity of Cro monomers for each other is high and virtually all Cro in the cell exists as dimers. In the absence of repressor, Cro can bind to the three operator sites in O_R. Figure 1.11 shows that one Cro dimer binds along the face of the helix at each site and is centered in exactly the same way as the repressor dimer. These two proteins—repressor and Cro—bind to the same three operator sites but play opposing roles in the switch mechanism.

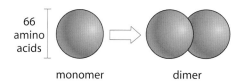

66 amino acids

monomer   dimer

Figure 1.10. Lambda's Cro. The Cro monomer contains only 66 amino acids, but folds into a globular structure about the same size as repressor's amino domain.

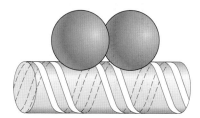

Figure 1.11. One Cro dimer bound to a 17 base pair operator site. Cro binds along the same face of the helix as does a repressor dimer.

THE ACTION OF REPRESSOR AND CRO

A key to understanding repressor's action is to consider the effect of attaching a single repressor dimer to O_R2, leaving O_R1 and O_R3 free. This scenario is not observed under normal circumstances, as we will explain, but it can be contrived experimentally, and the effects are revealing. The picture in Figure 1.12 shows that a repressor at O_R2 performs the two functions necessary for maintaining the lysogenic state: it turns off the *cro* gene *and* it turns on the repressor gene. We now consider the mechanism underlying each of these functions of repressor.

Negative Control

Repressor at O_R2 turns off the *cro* gene by preventing binding of RNA polymerase to P_R. It exerts this effect by covering part of the DNA that polymerase must see to bind to P_R. This principle of exclusion underlies many cases of negative control.

Positive Control

Repressor at O_R2 helps RNA polymerase bind and begin transcription at P_{RM}, the promoter governing transcription of *cl* in a lysogen. The increase in transcription is roughly tenfold. Repressor was named because of its ability to turn off all the other phage genes. Only later was it realized that repressor is a positive regulator as well, increasing transcription of its own gene.

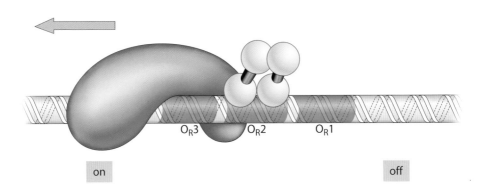

on   off

Figure 1.12. Repressor bound to O_R2. In a lysogen repressor is rarely bound to O_R2 unless it is also bound to O_R1. But if it were bound to O_R2 only, it would turn on P_{RM} and turn off P_R, so *cl* but not *cro* would be transcribed. Just as a protein-protein interaction holds two repressor monomers together in a dimer, so an interaction between repressor's amino domain and polymerase helps the enzyme bind and begin transcription at P_{RM}.

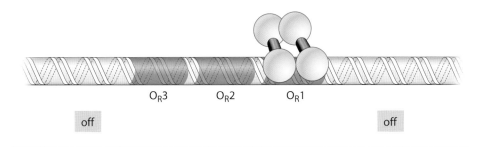

Figure 1.13. Repressor bound at O_R1. A repressor at O_R1 only would prevent polymerase from binding to P_R. P_{RM} would function only at a low (unstimulated) level because there is no repressor at O_R2 to help polymerase bind there.

The basis of repressor's action as a positive regulator is suggested in Figure 1.12: the repressor at O_R2 touches RNA polymerase at P_{RM}. The repressor dimer bound to O_R2 increases the affinity of P_{RM} for polymerase because polymerase is held there not only by contacts with DNA but also by a protein-protein contact with repressor.

A repressor dimer bound to O_R2, in sum, represses P_R by excluding binding of RNA polymerase to that promoter, but it encourages polymerase to begin transcription at P_{RM}. It prevents transcription to the right but aids transcription to the left. (We will see in Chapter Two that this difference is accounted for by the fact that O_R2 is slightly closer to P_R than to P_{RM}.)

We now consider the effects of a single repressor dimer bound either to O_R1 or to O_R3. As Figure 1.13 shows, a repressor at O_R1 would block binding of polymerase to P_R, but would be too far away to affect polymerase at P_{RM}. P_{RM} functions only at a very low level, and is labeled off in the figure, because no repressor is bound at O_R2. As Figure 1.14 shows, repressor at O_R3 would block binding of polymerase to P_{RM} and have no effect on P_R.

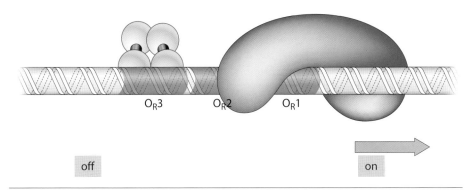

Figure 1.14. Repressor bound at O_R3. A repressor at O_R3 only would have no effect on P_R, which would be on. P_{RM} would be off.

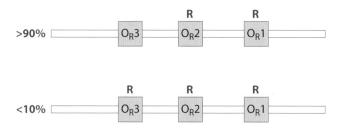

Figure 1.15. The λ operator in a lysogen. In a lysogen repressor dimers are bound primarily to sites O_R1 and O_R2, and occasionally to site O_R3 as well.

A series of snapshots of the inside of a λ-lysogen would reveal the situation summarized in Figure 1.15. Repressor dimers are virtually always bound at O_R1 and O_R2, but O_R3 is usually free of repressor. This arrangement of repressors at O_R turns off P_R but turns on P_{RM}: hence repressor, but not Cro, is synthesized. How is this pattern of repressor binding in a lysogen determined?

Cooperativity of Repressor Binding

Two factors determine the interaction of repressor dimers with the three operator sites. One is the affinity of the dimer for each of the three sites considered separately—we say that repressor has an "intrinsic" affinity for each site. The second factor is the interaction between repressor dimers bound to adjacent sites. Just as repressor at O_R2 helps polymerase bind to promoter P_{RM}, so repressors can interact with each other to facilitate binding.

Imagine a repressor dimer approaching a naked operator. Although this repressor might investigate all three sites, it would usually fix itself as shown in Figure 1.16 to O_R1—this is a way of saying that of the three sites, O_R1 has the highest affinity for repressor. This binding immediately increases the affinity of O_R2 for a second repressor dimer, because the second dimer not only contacts O_R2, but it also touches the previously bound repressor. The result of this interaction between repressors at O_R1 and O_R2 is that these two sites fill virtually simultaneously.

Figure 1.16 also shows that at higher repressor concentration O_R3, as well as O_R1 and O_R2, are filled. This binding of repressor to O_R3 turns off P_{RM}, as we have described. Site O_R3 binds repressor more weakly than does O_R2, despite the fact that the intrinsic affinities of these two sites for repressor are about the same and are about tenfold weaker than that of O_R1. Repressor binding at O_R2 is facilitated by the interaction with another repressor at O_R1, but the repressor at O_R3 must bind independently.

Why does repressor at O_R2 help polymerase bind to P_{RM}, but not help another repressor to bind to O_R3? Figure 1.16, and in greater detail Figure 1.17, shows the answer. Repressor-repressor interaction involves contacts between the carboxyl

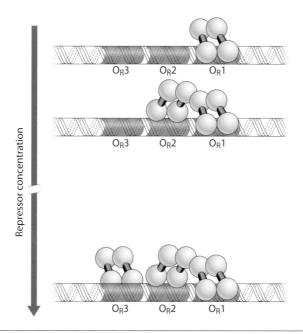

Figure 1.16. Repressor binding to the three sites in O_R. O_R1 binds repressor about ten times more tightly than does O_R2 or O_R3, so repressor first binds to O_R1. A second repressor very quickly binds to O_R2, but O_R3 continues to bind weakly, and is filled only at higher repressor concentration.

domains of adjacent dimers. Once a repressor dimer at O_R2 interacts with another dimer at O_R1, it is no longer available to interact with a dimer at O_R3.

We imagine that interaction between dimers at O_R1 and O_R2 requires that the proteins "lean" toward one another, a contortion that is allowed because of flexibility of the peptide connecting the amino and carboxyl domains. A dimer at O_R2 "leaning" to the right in Figure 1.16, and contacting a dimer at O_R1, cannot simultaneously contact a dimer at O_R3. Therefore O_R3 must fill independently. Repressor at O_R2 can interact with polymerase at P_{RM}, however, because that contact is made with repressor's amino domain, the domain that also contacts DNA. One of these amino domains is positioned just so this interaction occurs.

If we have correctly described how repressor dimers at O_R1 and O_R2 interact, we might expect that a repressor at O_R2 could "lean" to the left and interact with another at O_R3 if no repressor were bound at O_R1. Indeed, interaction between dimers at O_R2 and O_R3 occurs in the special circumstance that O_R1 is mutated or deleted so that no repressor can bind there. In that case, as shown in Figure 1.18, repressor dimers fill sites O_R2 and O_R3 simultaneously.

The protein-protein interactions we have been describing are examples of cooperativity. For example, we say that repressors bind cooperatively to O_R1 and O_R2. Because repressor dimers at O_R1 and O_R2 interact, *or* repressor dimers at O_R2 and O_R3 interact, we say the cooperativity is "alternate pairwise."

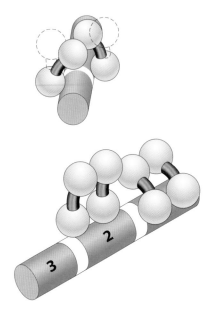

Figure 1.17. Interaction between adjacent repressor dimers. The linker between the amino and carboxyl domains of repressor is flexible, so a repressor at O_R2 can contact another at O_R1.

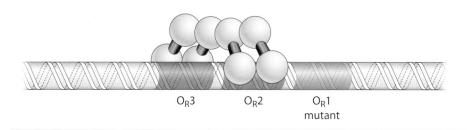

O_R3 O_R2 O_R1
mutant

Figure 1.18. Interaction between repressor dimers bound at O_R2 and O_R3. If O_R1 is mutant so that no repressor binds there, repressor at O_R2 is free to interact with another at O_R3. This interaction increases the repressor affinity of O_R2 and O_R3 about fivefold above their intrinsic affinities.

INDUCTION—FLIPPING THE SWITCH

In a lysogen, repressor bound at O_R1 and O_R2 keeps *cro* off while it stimulates transcription of its own gene *cl*, as summarized in Figure 1.19. Repressor is constantly being synthesized as the cells grow and divide, whereas the *cro* gene remains silent. If the repressor concentration increases—as might happen, for example, were cell division temporarily inhibited—repressor would bind also to O_R3 to turn its own gene off. As the cell began dividing again and the repressor concentration dropped to the proper level the *cl* gene would resume functioning,

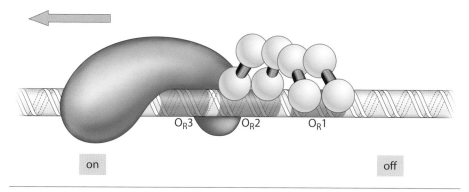

O_R3 O_R2 O_R1

on   off

Figure 1.19. Repressor and RNA polymerase in a lysogen. In a lysogen, repressor bound at O_R1 and O_R2 stimulates P_{RM} while simultaneously turning off P_R.

providing more repressor. Thus a constant level of repressor is maintained in the cell, despite fluctuations in growth rate.

In the cell, repressor continuously falls off the operator, only to rebind or to be replaced by another repressor molecule that happens to be nearby. The concentration of repressor is high enough to ensure that, at any given instant, O_R1 and O_R2 are very likely to be filled. Thus, in the absence of an inducing agent the lysogenic state is stable virtually indefinitely.

We are now in a position to follow the dramatic effects of ultraviolet light in a lysogen, beginning with the events illustrated in Figure 1.20. The primary effect of

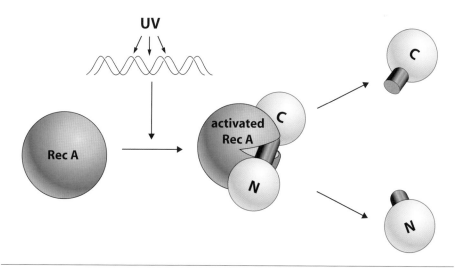

Figure 1.20. RecA cleavage of repressor. Repressor is cleaved between the amino acids alanine and glycine located in the linker between repressor's domains.

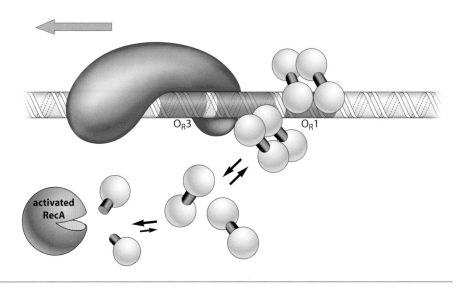

Figure 1.21. Repressor cleavage and induction. Cleaved repressors cannot dimerize, and so, following irradiation, when repressors fall off the operator they are not replaced.

inducers such as ultraviolet light and other inducing agents is to damage DNA. In a manner not fully understood, this damage leads to a remarkable change in behavior of a bacterial protein called RecA.

Under normal circumstances, RecA catalyzes recombination between DNA molecules, but when DNA is damaged, this protein also becomes a highly specific protease that cleaves λ repressor monomers. (RecA also cleaves other repressors, thereby turning on genes that help non-lysogenic cells survive the otherwise lethal effects of ultraviolet light, as described in Chapter Three.)

Figure 1.20 shows that the cleavage occurs at a specific site located in the region of repressor connecting the amino and carboxyl domains. The separation of the amino from the carboxyl domain effectively inactivates repressor, because the separated amino domains cannot dimerize, and their affinity for the operator in the monomeric form is too low to result in efficient binding at the concentrations found in a lysogen. Now as repressor dimers fall off the operator there are too few dimers to replace them, as illustrated in Figure 1.21.

Two changes result. First, as repressor vacates O_R1 and O_R2 the rate of repressor synthesis drops (because repressor is required to turn on transcription of its own gene); and second, polymerase binds to P_R to begin transcription of *cro*.

Cro's action is less complex than that of repressor. As Figure 1.22 shows, Cro dimers bind independently (non-cooperatively) to the three sites in O_R. Unlike repressor, Cro is strictly a negative regulator. A key to Cro's action is that the order

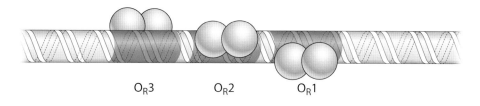

Figure 1.22. Cro bound to O_R. Cro dimers bind independently to each site in the tripartite operator.

of its affinity for the three sites in O_R is *opposite* to that of repressor. Figure 1.23 shows that the first Cro to be synthesized binds to O_R3. This prevents polymerase from binding to P_{RM} and abolishes further synthesis of repressor. At this point the switch has been thrown and lytic growth ensues.

As P_R continues to function and *cro* is transcribed, so are genes to the right of *cro*, whose products are needed for the early stages of lytic growth (see Chapter Three). More Cro is made until it reaches a level at which O_R1 and O_R2 are also filled and polymerase is prevented from binding to P_R, as shown in Figure 1.24. Cro, therefore, first turns off repressor synthesis and then, slightly later, turns off (or down) expression of its own and other early lytic genes.

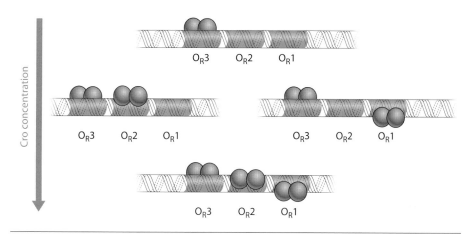

Figure 1.23. Order of binding of Cro dimers for sites in O_R. The affinity of site O_R3 for Cro is about tenfold higher than that of O_R2 or O_R1. After the first Cro dimer has filled O_R3, the second dimer binds to either O_R1 or to O_R2. The order with which Cro fills the sites is opposite to that with which repressor fills the sites. Thus on a wild-type O_R, the affinity for Cro is $O_R3 > O_R1 = O_R2$, whereas that for repressor is $O_R1 > O_R2 = O_R3$.

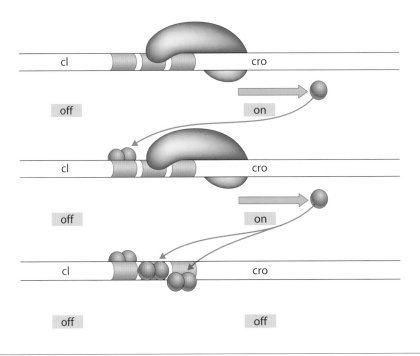

Figure 1.24. Binding Cro to OR3 blocks synthesis of repressor and binding to OR1 and OR2 turns down expression of its own gene.

COOPERATIVITY—SWITCH STABILITY AND SENSITIVITY

We have described three forms of cooperativity involving protein-protein interactions that contribute to making the switch mechanism highly efficient:

- Repressor monomers form dimers, the DNA-binding form of repressor. These dimers freely dissociate into monomers, and in the cell monomers and dimers are in equilibrium. One way to describe the binding of a dimer is to say that one monomer helps the other to bind the operator, that is, two monomers bind cooperatively to an operator site;

- Repressor dimers bind cooperatively to adjacent sites in the operator. The predominant effect is that a repressor dimer at O_R1 helps another bind to O_R2;

- Repressor at O_R2 helps polymerase bind and begin transcription at P_{RM}.

The sum of these effects is best grasped by referring to Figure 1.25 which plots, approximately, the activity of P_R versus the concentration of repressor. Note that in a lysogen enough repressor is synthesized to repress P_R about 1000-fold. Over the

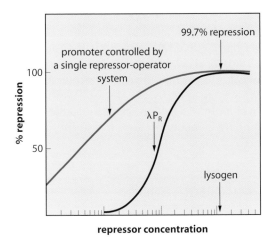

Figure 1.25. Repression as a function of repressor concentration in two systems. By following the black line from right to left we see that repression is maintained in a lysogen only until the repressor concentration drops about fivefold. The system then responds dramatically to any further drop in repressor concentration, and induction occurs. In contrast, a single-site operator-repressor interaction (blue line) would react much more sluggishly to a change in repressor concentration.

first twofold or threefold drop in repressor concentration from this high level, the activity of P_R remains unchanged. In effect, repression is buffered against ordinary fluctuations in repressor concentration, so that lysogens are rarely "accidentally" induced. But when the repressor concentration has dropped about fivefold, P_R responds dramatically, functioning at about 50% of its fully unrepressed level. This allows synthesis of Cro and of other lytic gene products, thereby flipping the switch.

It is instructive to consider a hypothetical switch devoid of cooperativity. For example, if O_R contained only a single site corresponding to O_R2, the system might work, but only crudely. If this site bound a repressor dimer tightly enough to ensure 1000-fold repression in a lysogen, induction would occur only very inefficiently. Over 99% of the repressor would have to be inactivated to trigger lytic growth, and this is a difficult requirement to meet.

The properties of λ repressor thus provide an answer to a problem of concern to developmental biologists: how a relatively mild change in the concentration of a regulatory protein can reliably cause a switch in gene expression. Cooperativity between repressor monomers—that is, binding of a dimer to each site—and between repressor dimers on the DNA magnifies greatly the effect of a drop in concentration of repressor monomers. Moreover, the fact that repressor stimulates transcription of its own gene, acting as a positive regulator, means that synthesis of new repressor must decrease as the repressor concentration falls. When the repressor concentration drops to the critical point, lytic growth ensues.

The two curves of Figure 1.25, one describing the cooperative λ repressor system, the other a single-site noncooperative system, are similar to the curves that describe the behavior of the oxygen-carrying molecules hemoglobin and myoglobin. Hemoglobin carries oxygen from lungs to tissues, and myoglobin helps move oxygen within muscle. The oxygen pressure in tissues is only about fivefold lower than that in lungs; nevertheless, hemoglobin efficiently accepts oxygen in the lungs and releases it in the tissues. The curve describing the binding of oxygen to hemoglobin resembles the highly cooperative black curve of Figure 1.25. The four subunits of hemoglobin, each of which binds one oxygen molecule, bind oxygen cooperatively, and this binding is highly sensitive to the oxygen concentration. A non-cooperative curve, similar in shape to that drawn in red in Figure 1.25, describes the binding of oxygen to myoglobin. Each myoglobin molecule binds only one oxygen molecule, and this binding is much less sensitive to the oxygen concentration.

THE EFFECT OF AUTOREGULATION

We have noted that, in addition to activating transcription of its own gene in lysogens (by binding to O_R1 and O_R2) repressor uses O_R3 to limit its own concentration. Through this interaction at O_R3, repressor ensures that its concentration never exceeds the level at which the prophage can respond efficiently to the inducing signal. Too much repressor would inhibit induction in two ways.

- RecA cleaves repressor monomers only rather slowly; if the repressor concentration were too high, the fraction of repressor cleaved would not be sufficient to cause induction.

- Even if repressor at high concentrations were cleaved, repression would not be lifted. The reason is that even though cleaved repressor monomers cannot form dimers, the severed amino domains can still bind to the operator non-cooperatively. At high enough concentrations, these amino domains will fill the three binding sites whether or not they are part of repressor dimers.

    Negative and positive self-control circuits—as exemplified by the two aspects of repressor's self-regulation—respond quite differently to perturbations. Negative control is self-correcting, or homeostatic, whereas positive control is destabilizing. If only negative control were operative, should the repressor concentration increase or decrease, the rate of repressor synthesis would decrease or increase so as to bring the concentration back to the equilibrium level. In contrast, a positive control circuit would respond by exacerbating any change in repressor level—an increasing repressor concentration would increase the rate of repressor synthesis and a falling concentration would decrease that rate.

OTHER CASES

We know of many examples of phages other than λ that, in addition to growing lytically, form inducible lysogens of their bacterial hosts. Do the switches in these cases resemble the λ switch? Consider two other phages—434, which grows on *E. coli,* and P22, which grows on *Salmonella typhimurium*. 434- and P22-lysogens, like λ-lysogens, are efficiently induced by ultraviolet light. Both phages encode a repressor and a Cro, which act on a region of their own DNA analogous to λ's O_R.

Despite differences in detail, the following description holds for all three phages: the right operator contains three repressor binding sites, two of which, O_R1 and O_R2, are filled by repressor in a lysogen. This state depends on interactions between adjacent repressor dimers. Repressor monomers have two structural domains and are in equilibrium with dimers, the DNA-binding form. Repressor dimers occupying sites O_R1 and O_R2 turn off transcription from P_R and turn on transcription from P_{RM}. Upon ultraviolet induction, the repressor is cleaved and thereby inactivated, and the first protein synthesized from P_R is Cro. Cro binds first to O_R3 to turn off repressor synthesis and later binds to O_R1 and O_R2 to decrease early lytic transcription. The fact that these features are widespread supports the view that they are fundamental components of the switch.

FURTHER READING: RELATED REVIEWS

1. Gussin, G., Johnson, A., Pabo, C., and Sauer, R. (1983). Repressor and Cro protein: structure, function, and role in lysogenization. In *Lambda II* , R.W. Hendrix, J.W. Roberts, F.W. Stahl, and R. Weisberg, eds.) New York: Cold Spring Harbor), pp. 93–123.

2. Johnson, A.D., Poteete, A.R., Lauer, G., Sauer, R.T., Ackers, G.K., and Ptashne, M. (1981). λ repressor and *cro*-components of an efficient molecular switch. *Nature* 294, 217–233.

3. Lwoff, A. (1953). Lysogeny. *Bacteriol. Rev.* 17, 269.

4. Ptashne, M. (1984). Repressors. *Trends Biochem. Sci.* 9, 142–145.

5. Ptashne, M., Backman, K., Humayun, M.Z., Jeffrey, A., Maurer, R., Meyer, B., and Sauer, R.T. (1976). Autoregulation and function of a repressor in bacteriophage λ. *Science* 194, 156–161.

6. Ptashne, M. and Gilbert, W. (1970). Genetic repressors. *Sci. Am.* 222, 36–44.

7. Ptashne, M., Jeffrey, A., Johnson, A.D., Maurer, R., Meyer, B.J., Pabo, C.O., Roberts, T.M., and Sauer, R.T. (1980). How the λ repressor and Cro work. *Cell* 19, 1–11.

8. Roberts, J. and Devoret, R. (1983). Lysogenic induction. In *Lambda II* R.W. Hendrix, J.W. Roberts, F.W. Stahl, and R. Weisberg, eds. (New York: Cold Spring Harbor), pp. 123–145.

CHAPTER TWO

PROTEIN-DNA INTERACTIONS AND GENE CONTROL

Regulatory proteins—λ's repressor and Cro, for example—bind tightly to specific DNA sequences 15–20 base pairs long. Each of these proteins must select its operator site from among the five million or so base pairs of DNA in a bacterium. This chapter examines the structures of these proteins to show how they accomplish this feat. The principle that emerges is simple: the structure of the protein is complementary to the DNA structure—if the DNA sequence is correct the protein and DNA molecules fit together like lock and key. We also show how a regulatory protein, once bound, can control gene expression negatively or positively.

THE OPERATOR

The drawing of Figure 2.1 reveals how, in principle, a sequence of double-helical DNA can be recognized. Edges of the base pairs are exposed in the major and minor grooves that run along the helix. Each base pair (A:T, T:A, G:C, C:G) exposes a different pattern of chemical groups that a protein might recognize. These groups are not the ones involved in complementary base pairing—those groups become accessible only if the strands separate, as happens during replication or transiently during transcription. We will see that λ repressor and Cro bear protuberances that penetrate the major groove to read the DNA base sequence.

Each of the operator sites recognized by repressor and Cro is nearly—but not exactly—symmetric. To understand this, consider the perfectly symmetric sequence of Figure 2.2. If this segment of DNA were rotated 180° in the plane of the page and around the dot at its center, the identical molecule would be generated. The distinctive feature of this particular segment of double-stranded DNA is that its base sequence remains unchanged by the rotation.

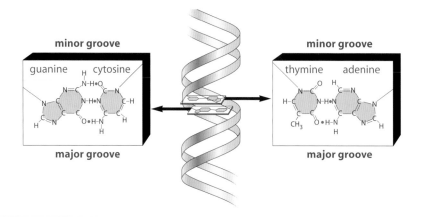

Figure 2.1. A segment of DNA. Peering into the major and minor grooves of DNA we see that each base pair can be identified by characteristic chemical groups that lie along the edges of the base pair. The DNA need not be unwound to read its sequence.

Another way to describe the symmetry of the sequence of Figure 2.2 is to imagine a tiny demon standing at the middle of the molecule, the position represented by the dot. The demon, facing right and then left, would see identical corridors of chemical groups. We say that our sequence is twofold rotationally symmetric. Were one base pair to be changed the sequence would be nearly, but not exactly, symmetric.

The sequences recognized by λ repressor and Cro are listed in Table 2.1. Three of these 17 base pair sites are from λ's right operator, O_R, and three from the left operator, O_L. The six sites differ somewhat in sequence, and they do not all have identical affinities for repressor and Cro.

None of the λ operator sites is perfectly symmetric, and if we were to examine only one of them—O_R2, for example—the symmetry would not be particularly striking. But if we split each operator into two half-sites, and align the half-sites as shown in Table 2.2, a clear pattern emerges. The table lists the frequency with which various bases are represented at each position. The consensus half-site, written in double-stranded form, is:

TATCACCGC
ATAGTGGC

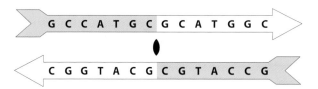

Figure 2.2. A symmetric DNA sequence. The sequence of the top strand, read left to right, is the same as that of the bottom strand read right to left. The black diamond indicates the axis of symmetry.

Table 2.1. The six operator sites recognized by λ repressor and Cro.

O_L1	T	A	T	C	A	C	C	G	C	C	A	G	T	G	G	T	A
	A	T	A	G	T	G	G	C	G	G	T	C	A	C	C	A	T
O_R1	T	A	T	C	A	C	C	G	C	C	A	G	A	G	G	T	A
	A	T	A	G	T	G	G	C	G	G	T	C	T	C	C	A	T
O_L2	T	A	T	C	T	C	T	G	G	C	G	G	T	G	T	T	G
	A	T	A	G	A	G	A	C	C	G	C	C	A	C	A	A	C
O_L3	T	A	T	C	A	C	C	G	C	A	G	A	T	G	G	T	T
	A	T	A	G	T	G	G	C	G	T	C	T	A	C	C	A	A
O_R2	T	A	A	C	A	C	C	G	T	G	C	G	T	G	T	T	G
	A	T	T	G	T	G	G	C	A	C	G	C	A	C	A	A	C
O_R3	T	A	T	C	A	C	C	G	C	A	A	G	G	G	A	T	A
	A	T	A	G	T	G	G	C	G	T	T	C	C	C	T	A	T

The sites are listed in the order of their intrinsic affinities for a λ repressor dimer. The central base pair, the axis of symmetry, is shown in blue.

Table 2.2. The 12 half-sites of the λ operators.

Operator site	1	2	3	4	5	6	7	8	9 base position
O_L1	T	A	T	C	A	C	C	G	C
	T	A	C	C	A	C	T	G	
O_R1	T	A	T	C	A	C	C	G	C
	T	A	C	C	T	C	T	G	
O_L2	T	A	T	C	T	C	T	G	
	C	A	A	C	A	C	C	G	C
O_L3	T	A	T	C	A	C	C	G	C
	A	A	C	C	A	T	C	T	
O_R2	T	A	A	C	A	C	C	G	T
	C	A	A	C	A	C	G	C	
O_R3	T	A	T	C	A	C	C	G	C
	T	A	T	C	C	C	T	T	
Consensus	T_9	A_{12}	T_6	C_{12}	A_9	C_{11}	C_7	G_9	C_5
	C_2		C_3		T_2	T_1	T_4	T_2	T_1
	A_1		A_3		C_1			G_1	C_1

Each of the six full sites listed in Table 2.1 is represented here as two half-sites. Only one strand is written for each half-site. The upper strand corresponds to the left half-site of the top strand of Table 2.1 reading left to right; the lower strand corresponds to the bottom strand of the right half-site of Table 2.1, reading right to left. The positions in the operator are numbered at the top. At the bottom the frequency with which each base appears is given for each position. One of two bases could be chosen to represent position 9 for each case.

Each λ operator site is more or less closely related to the following symmetric sequence. It contains two consensus half-sites and an axis of symmetry that runs through the central base pair:

TATCACCGCCGGTGATA
ATAGTGGCCGCCACTATT

The sequence of bases at each operator site must provide a pattern of functional groups recognized by repressor and Cro. Each of the proteins must also be able to distinguish between the sites, binding with the proper affinity order as discussed in Chapter One. We examine next the structures on repressor and Cro that recognize these sequences.

REPRESSOR

In the Introduction we noted that proteins fold into characteristic shapes determined by their amino acid sequences. Although different amino acid sequences usually fold into different overall shapes, certain smaller structural motifs are found in many proteins. One common motif is the α-helix.

As shown in Figure 2.3, the α-helix is formed by the spiraling of a single chain of amino acids. An important difference between the α-helix and DNA's double helix is that, in the protein structure, the backbone is on the inside and the characteristic groups of each residue—the amino acid side chains—are on the outside.

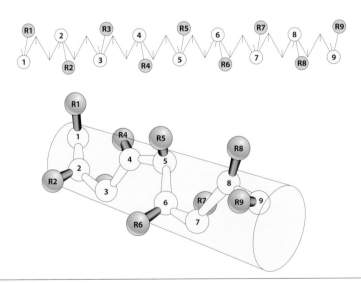

Figure 2.3. A chain of amino acids unfolded and in the form of an α-helix. The side chains R_1, R_2, etc. are different for each of the 20 different amino acids. In the α-helix, these side chains protrude from the backbone which is shown here enclosed in a barrel. One turn of the α-helix comprises 3.6 amino acid residues.

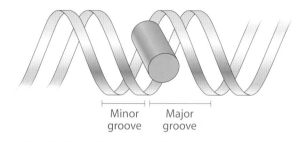

Minor Major
groove groove

Figure 2.4. An α-helix in a major groove. The side chains that protrude from the α-helix, not shown here, would extend to the extremities of the DNA major groove.

Figure 2.4 shows how an α-helix fits neatly into the major groove of DNA. The functional groups along a surface of the α-helix are positioned to interact with—touch—chemical groups on the edges of the DNA base pairs.

To understand how λ repressor uses an α-helix to recognize its operator we must consider the structure of repressor in greater detail than our dumbbell representation in Chapter One. As shown in Figure 2.5, repressor's amino domain is folded into five successive stretches of α-helix. Alpha-helix 3 lies exposed along the surface of the molecule—this is repressor's "recognition" helix.

In the repressor dimer, the recognition helices—one on each monomer—are separated by the same distance that separates successive segments of the major

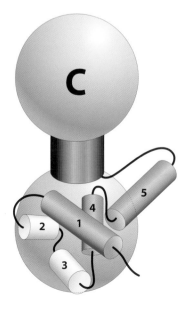

Figure 2.5. Lambda repressor. The five α-helices that comprise repressor's amino domain are connected by segments of the amino acid chain. Helix 1 is near the very amino terminus of the protein. The structures of the linker and of the carboxyl domain (C) are not known.

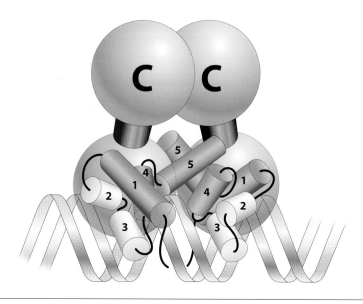

Figure 2.6. Lambda repressor bound to an operator site. A pair of repressor amino domains fits on a 17 base pair operator site.

groove along one face of the DNA. Figure 2.6 shows that when the dimer docks with DNA, each recognition helix fits into the major groove. Thus the symmetry of the protein matches that of the DNA when the dimer is positioned on the operator.

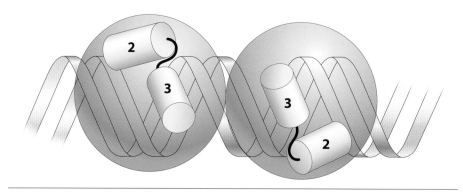

Figure 2.7. Bihelical units on the operator. Two bihelical units are symmetrically positioned on an operator site. Bihelical units are also called helix-turn-helix motifs.

Figure 2.8. The arms of λ repressor. Seven amino acids extend from the end of α-helix 1. By wrapping around the DNA they contact bases near the center of the operator on its backside. Note that the DNA has been turned around, compared with the previous figure, to illustrate its backside.

A second α-helix—helix 2—is highlighted in addition to the recognition helix in Figures 2.5 and 2.6. Figure 2.7 shows that this α-helix lies across but not in the major groove. Helix 2 helps to position helix 3 in the major groove of the DNA. We refer to these two α-helices as a bihelical unit or structure. As we shall see, many regulatory proteins use a similar pair of α-helices in binding to DNA.

Now we see, in principle, how repressor binds selectively to its operator. Only if there is a match between the amino acid side chains along repressor's recognition helix—its α-helix 3—and DNA functional groups exposed in the major groove will the protein bind tightly. To understand the role of symmetry in the binding, recall our symmetry-detecting demon. Were the demon to walk along helix 3 in either monomer, from amino to carboxyl end—that is, toward helix 4—it would see the same (or nearly the same) succession of protein-DNA interactions. (The reason we say "nearly the same" is that the sequence of each operator site is not perfectly symmetric.)

In addition to penetrating the DNA with its recognition α-helix, λ repressor embraces the DNA with a pair of flexible arms that extend from its amino end. As the repressor binds the operator, its arms wrap around and make specific contacts on the backside of the DNA in the major groove. Many regulatory proteins use recognition α-helices to recognize specific base sequences, but the arms of λ repressor shown in Figure 2.8 may be an example of a less widely used sequence-reading device.

CRO

Figure 2.9 shows Cro's structure. It includes three α-helices and, in addition, three regions that form so-called β-sheets. (The β-sheet is a second common structural

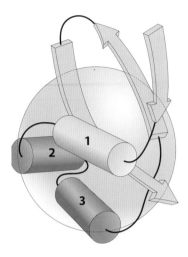

Figure 2.9. Cro. The three flattened arrows represent the segments of a β-sheet that, in addition to its three α-helices, form Cro's structure.

motif found in many proteins. Our pictures represent β-sheets as flattened arrows to distinguish them from the α-helical barrels.)

Cro's helix 3—just as with repressor—is its recognition helix. The relative spatial orientation of Cro's α-helices 2 and 3 is virtually identical with that of the corresponding pair of helices in repressor. It is remarkable to find two α-helices positioned so similarly to form a bihelical unit in different proteins. In the Cro dimer, the symmetrically related recognition helices—one on each monomer—fit into successive segments of the major groove as shown in Figure 2.10.

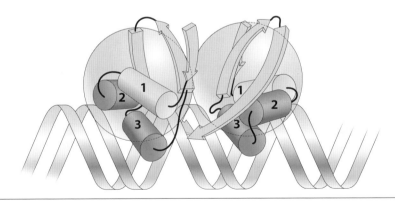

Figure 2.10. Cro bound to an operator site. A Cro dimer docks with DNA in much the same way as does a repressor dimer.

AMINO ACID-BASE PAIR INTERACTIONS

The amino acid sequence comprising the recognition helix of repressor and Cro are for the most part different as shown in Figure 2.11. This is not surprising. In each case, one surface (the "inside") of the helix fits against the body of the protein, and the "outside" surface fits into the DNA's major groove to make specific contacts. The inside surfaces of the recognition helices differ because the bodies of the proteins with which they interact are different in repressor and Cro. The sequences along the outsides of the recognition helices are similar but not identical— although repressor and Cro bind to the same operator sites, they do so with different relative affinities.

Figure 2.11 also shows the pattern of interactions between amino acids in the recognition helices of repressor and Cro with bases in two operator half-sites. The pattern suggests how repressor and Cro recognize the same operator sites, but distinguish between them, repressor preferring O_R1 to O_R3 and vice versa for Cro.

Both recognition helices begin with the sequence glutamine-serine (Gln-Ser) and then diverge in sequence. Both proteins use Gln at position 1 and Ser at position 2 to contact positions 2 and 4, respectively, of the operator. These two positions in the operator are just those, as shown in Table 2.2, whose identities are invariant in all the operator half-sites. Thus, repressor and Cro use identical amino acids to contact bases that are identical in all the λ operator sites.

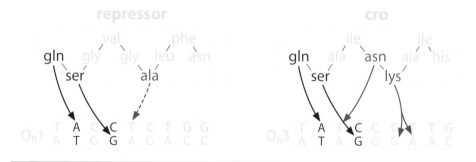

Figure 2.11A. Pattern of amino acid-base pair interactions. To see how the individual units of these recognition helices are oriented and numbered, compare with Figure 2.3 and Figure 2.11B. In each of the proteins the side chains (R groups) of amino acids 1, 2, 5, and 6 point toward the DNA, and those of residues 4 and 7 point toward the body of the protein. The arrow connecting the Ala repressor to the T:A base pair at position 5 is dashed because, although the presence of the Ala causes a preference for that T:A, the basis for this preference is not known. The figure omits three additional interactions: a Lys residue in λ repressor's arm and an Asn residue just beyond the end of repressor's helix 3 both contact the C:G found at position 6 of both O_R1 and O_R3. And, at position 8, repressor's arm, which is not shown, prefers G:C to T:A. Only one half-site is shown for each operator. In both cases, the other half-site is the consensus sequence of Table 2.2.

λ repressor   λ Cro

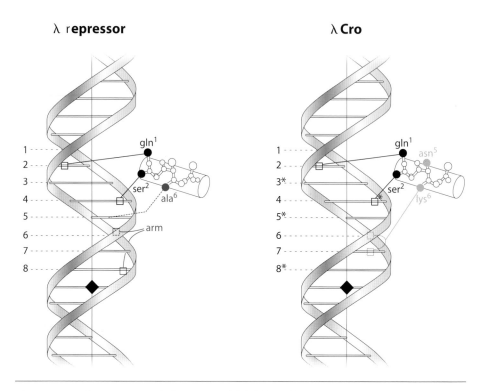

Figure 2.11B. The recognition helices are unfolded from their position in the major groove of DNA. Note the correspondence between the positions of the protruding amino acid side chains and the contacted bases. The bases in the operator are numbered as in Table 2.2, and the three base pairs that distinguish O_R1 from O_R3 are starred. The diamond denotes the center of symmetry of the operator.

Amino acids other than the conserved Gln-Ser pair enable the two proteins to distinguish between the individual operator sites. Thus, for example, asparagine (Asn) in Cro's recognition helix contacts position 3 in the operator, preferring the base pair found in the non-consensus half-site of O_R3 to that found in the corresponding position in O_R1. The arm of λ repressor, which contacts positions 6 and 8 in the operator, also helps repressor to distinguish between the sites.

Figure 2.12 shows in detail one of the amino acid-base interactions indicated in Figure 2.11. Thus, Gln makes two bonds to positions on the base adenine (A) exposed in the major groove.

Figure 2.13 shows the identities of certain amino acids at corresponding positions of the bihelical structure that are the same or are chemically similar ("conserved") in repressor and Cro. (These conserved amino acids are not those involved in recognition of specific sequences.) Three of these amino acids are in the elbow between the two helices, and two, connected by a line in the figure, lie one in each

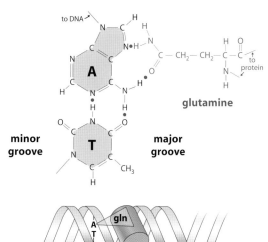

glutamine

minor groove

major groove

Figure 2.12. An amino acid-base pair contact. Glutamine is shown contacting the base A in the major groove. Two hydrogen bonds are formed between the side chain of the amino acid and the edge of the base.

helix. These residues evidently serve to maintain the constant angle between the two helices. A sixth conserved residue is found at the top of the helix preceding the recognition helix. This amino acid interacts with a phosphate in the backbone of one of the DNA strands, thereby helping to position the bihelical structure properly on the DNA.

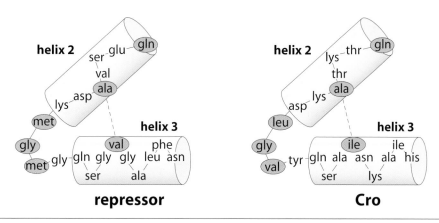

Figure 2.13. Conserved positions in the bihelical units of repressor and Cro. Repressor and Cro each bear an amino acid in helix 2 that interacts with another in helix 3—alanine and valine in the case of repressor, and alanine and isoleucine in the case of Cro. These bonds help position the two helices, as do the circled residues in the elbows. The side chain of methionine (Met) is chemically similar to those of leucine (Leu) and valine (Val).

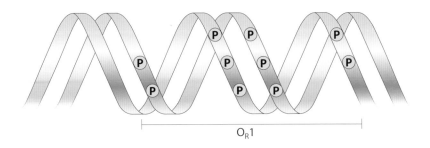

Figure 2.14. Phosphates in contact with repressor bound at $O_R 1$. The phosphates contacted by repressor lie along one face of the helix and are symmetric about a twofold axis through the center of the operator. Cro contacts a subset of these phosphates.

Many known and putative specific DNA-binding proteins—from an array of organisms—have a pattern of conserved residues identical or closely related to that found in λ repressor and Cro. In several cases other than λ repressor and Cro, this pattern of amino acids has proved diagnostic of a bihelical unit, one member of which is a recognition helix. The relevant homologies do not extend over the entire bihelical regions; rather they are confined to those few regions that—evidently—fix the spatial relation between the members of the bihelical unit.

Lambda repressor and Cro bind very tightly to their operators, a matter to which we return in Chapter Four and in Appendix 1. In our discussion thus far we have emphasized the determinants of specificity, namely, specific interactions between DNA functional groups and the amino acids in the recognition α-helix and, in the case of λ repressor, the end of its flexible arm. But some of the tightness of binding, probably most, comes from interactions between other parts of the protein and the DNA. For example, when λ repressor docks with its operator the protein interacts with phosphates that lie along the backbones of the DNA helix, shown in Figure 2.14. These and other tight interactions are allowed only if the specificity probes find the proper DNA groups with which to interact.

We are just beginning to decipher the interactions of amino acid chains with base pairs that allow specific binding. We suspect that in general no more than three or four amino acids on any particular protein monomer determine specificity. Does the recognition helix present itself identically to the DNA in every case so that amino acids at specified positions interact with bases at specified positions in the operator? Is there a simple code describing amino acid-base pair interactions?

We have treated DNA as a rigid rod but this is a simplification. We know, for example, that the precise parameters describing the helical DNA structure are determined in part by sequence. Will these sequence-specific alterations in structure influence protein-DNA interactions significantly? It is likely, for example, that bases not actually contacted by a bound protein might nevertheless influence the binding by subtly altering local DNA structure or flexibility. We do not yet know how important such factors are.

THE PROMOTER

The enzyme RNA polymerase in *E. coli* recognizes many promoters near the beginning of many different genes. Recall from the Introduction that a promoter spans about 60 base pairs, including 20 base pairs downstream (that is, in the direction it will travel) from where transcription begins. In general, no two promoter sequences are identical, but all have two characteristic blocks of sequence: one centered about 10 base pairs upstream of the RNA start site and the other about 35 base pairs upstream. By convention, the first DNA base copied, the RNA start site, is numbered +1, and so these conserved promoter sequences are centered at positions –10 and –35.

By comparing many sequences we can deduce a consensus promoter with the two blocks of sequence shown in Figure 2.15. Any given promoter has a sequence at these positions more or less clearly related to these conserved elements.

The sites of the promoter contacted by polymerase lie along one face of the helix from about position –10 to –40. Between –10 and +1 the DNA is opened so that one of the strands can be copied into mRNA. The structural features of polymerase that enable it to bind selectively to promoters are not known.

In general a promoter that has a good match to the consensus sequence in the –35 and –10 regions will work well. But if the sequence deviates greatly from the consensus, then an activator protein is usually required to help the polymerase bind and begin transcription efficiently.

Table 2.3 compares the consensus promoter sequence with that actually found at P_R and P_{RM}. P_R makes a better match at both the –35 region and the –10 region than does P_{RM}. This may be why RNA polymerase can bind efficiently to P_R and begin transcription without the aid of any regulatory protein, but requires an auxiliary protein—λ repressor bound at O_R2—if it is to bind efficiently and begin transcription at P_{RM}.

Figure 2.15. A consensus promoter. This promoter directs transcription rightward. The first base copied is at position +1, and –10 and –35 identify bases at those positions "upstream" of the transcription start. The conserved sequence around –10 is sometimes called the "TATA" box. The separation between the conserved elements (TATA and –35) varies in different promoters between 15 and 18 base pairs, with 17 base pairs being optimal.

Table 2.3. Lambda promoter sequences compared with the consensus promoter. The differences are shown in blue.

	−35		−10
Consensus	T T G A C A	– 17 bp –	T A T A A T
λP_{RM}	T A G A T A	– 17 bp –	T A G A T T
λP_{R}	T T G A C T	– 17 bp –	G A T A A T

GENE CONTROL

We are now in a position to understand how the promoter and operator sequences are arranged so that the effects of repressor and Cro, described in Chapter One, are realized.

Figure 2.16 shows the DNA sequence that extends from *cl* to *cro*. Compare with Figure 1.4, and note here how promoter sequences overlap (interdigitate with) operator sequences. As we saw in Chapter One (see Figure 1.13) when O_R1 is filled with repressor or with Cro, RNA polymerase is excluded from P_R because either bound protein covers part of the surface of the DNA helix that must be occupied by RNA polymerase. At O_R3 the same rule holds: either regulatory protein would block binding of RNA polymerase to P_{RM}. At O_R2 the situation is more delicate.

Recall from the previous chapter that O_R2 mediates both negative and positive control: repressor bound at this site turns off P_R, while it stimulates P_{RM}. The mechanism of negative control is the familiar one: repressor bound to O_R2 would cover part of the DNA surface that polymerase must see to bind P_R. The promoter surface overlapped by repressor at this site is less than if O_R1 were occupied, but even this degree of overlap suffices to effect repression.

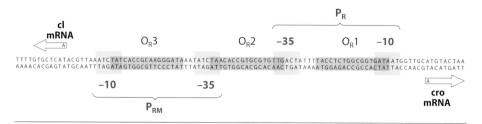

Figure 2.16. Linear relationship between promoter and operator sites around O_R. Some base pairs serve dual functions in the region between *cl* and *cro*. For example, three of the base pairs of O_R2 form part of the −35 region of P_R.

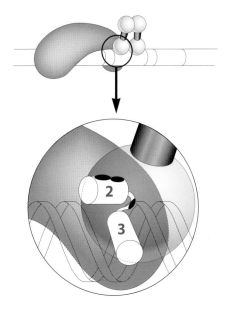

Figure 2.17. Lambda repressor as an activator of transcription. Repressor amino acids that interact with polymerase to mediate positive control are located along helix 2 and in the bend between helix 2 and helix 3.

How does repressor bound to O_R2 stimulate P_{RM}? Figure 2.16 shows that O_R2 is one base pair closer to P_R (counting from the transcription start site) than it is to P_{RM}. Because of this difference, repressor at O_R2 does not cover any part of the surface of P_{RM}. Rather, repressor at O_R2 closely approaches—touches—RNA polymerase bound at P_{RM}. As noted in Chapter One (see Figure 1.12), this interaction helps polymerase bind and begin transcription at P_{RM}.

The surface of repressor that touches polymerase and mediates positive control is indicated in Figure 2.17. The amino acids on this surface patch are specially configured to interact with polymerase and, if they are changed, the repressor can bind to O_R2 but can no longer stimulate P_{RM}. We imagine that were O_R2 positioned one base pair closer to P_{RM}—thereby mimicking its relation to P_R—repressor at O_R2 would block binding of polymerase to P_{RM} rather than help it.

Cro at O_R2 also blocks polymerase from binding to P_R. The region of P_R covered by Cro at O_R2 is the same as that covered by repressor at O_R2. If Cro is bound only to O_R2—an artificial situation that can be contrived—it fails to stimulate P_{RM}. The reason is that, although Cro might be positioned properly to contact polymerase at P_{RM}, it lacks the appropriate amino acids along the contacting surface that would favorably interact with polymerase.

It is worth emphasizing that contemplation of a linear representation of operator and promoter sequences can be misleading. For example, Figure 2.16 seems to suggest that O_R2 overlaps both P_R and P_{RM} such that repressor bound there would repress both promoters. The three-dimensional analysis makes clear the fact that proteins can recognize overlapping sequences but be bound on different surfaces of the DNA helix.

The analysis of this chapter suggests that a protein will repress a promoter if it covers some surface of the DNA helix to which polymerase must bind. There are various ways that operator and promoter sequences can be arranged to ensure repression, but in each case, the sequences must overlap (that is, interdigitate) so that each maintains its function.

Positive control requires an interaction between the bound regulatory protein and polymerase. The regulatory protein must have appropriate amino acids on its surface in a position to bind to polymerase. The surface that interacts with polymerase is different from the surface that binds DNA. The functions—DNA-binding and positive control—can be distinguished: two proteins may bind identically while only one manifests positive control.

FURTHER READING: RELATED REVIEWS

1. Hawley, D.K. and McClure, W.R. (1983). Compilation and analysis of *Escherichia coli* promoter DNA sequences. *Nucl. Acids Res.* 11, 2237–2255.

2. Maniatis, T. and Ptashne, M. (1976). A DNA operator-repressor system. *Sci. Amer.* 234, 64–76.

3. Pabo, C.O. and Sauer, R.T. (1984). Protein-DNA recognition. *Ann. Rev. Biochem.* 53, 293–321.

4. Ptashne, M., Johnson, A.D., and Pabo, C.O. (1982). A genetic switch in a bacterial virus. *Sci. Amer.* 247, 128–140.

5. Siebenlist, U., Simpson, R.B., and Gilbert, W. (1980). *Escherichia coli* RNA polymerase interacts homologously with two different promoters. *Cell* 20, 269–281.

CHAPTER THREE

CONTROL CIRCUITS—
SETTING THE SWITCH

M any viruses grow in only one way. Soon after infecting the host cell the viral genes work vigorously—new proteins are synthesized that extensively replicate the viral chromosomes, package them in new viral particles, and lyse the cell.

As we have noted, λ multiplies in such a lytic fashion, but it also has an alternative. In a lysogenic bacterium the phage genes required for lytic growth are turned off and the phage chromosome is replicated passively by proteins encoded and synthesized by the bacterium. In the parlance of Chapter One we say that the master control element—the switch—has been set so that a single phage protein, the repressor, dominates. Recall that the repressor turns off the other phage genes as it turns on its own gene, *cI*. This otherwise stable situation is perturbed by inducing agents, such as ultraviolet light, that flip the switch by destroying repressor, and lytic growth begins.

This chapter is concerned primarily with two questions concerning gene regulation in λ.

- How does λ "decide" whether to grow lytically or to lysogenize a newly infected bacterium? Presumably it is a useful trick to multiply surreptitiously, as part of a lysogen, if the conditions for vigorous lytic growth are not optimal.

- How does λ regulate its genes as growth proceeds down either of its two separate pathways?

When the phage grows lytically following infection it must, in an orderly fashion, replicate and package its DNA and then lyse the cell, all the while preventing synthesis of repressor. When a lysogen is induced, a further task is faced by the prophage about to begin lytic growth: an enzyme must be synthesized that releases—excises—the prophage from the host chromosome.

In contrast, when the phage begins the lysogenization process it must synthesize the enzymes that integrate its chromosome into that of the host and begin synthesis of repressor, while preventing expression of the various lytic genes. In other words, the switch of Chapter One must be set in the lysogenic position.

The lysis-lysogeny decision is an instructive example of how the environment can influence the choice of developmental pathways. As we shall see, the first few steps of gene regulation that occur upon λ infection are identical whether the phage is ultimately to lyse the cell or to lysogenize it. At the critical step the state of the host is sensed by a phage regulatory protein, and subsequent events are appropriately funneled down one of the two pathways.

We shall see that whether λ is growing lytically or is establishing lysogeny, its pattern of gene regulation is organized as a *cascade:* one regulatory protein typically turns on (or off) a block of genes; that block of genes typically includes another regulatory gene whose product in turn activates (or represses) a second block of genes, and so on.

We begin with a brief overview of the activities of an infecting λ chromosome as it travels down the lytic or the lysogenic path. We then examine the patterns of gene expression in more detail, and return to the question of how the lysis or lysogeny decision is made.

A BRIEF OVERVIEW OF λ GROWTH

The Genetic Map

A simplified map of the λ chromosome is shown in Figure 3.1. Only six genes are named individually in this representation: the regulatory genes *cl, cll, clll, N, cro,* and *Q.* The remaining genes are indicated in groups according to the functions of the proteins they encode.

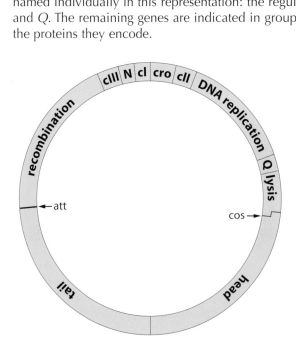

Figure 3.1. The λ chromosome. In general, genes of related function are grouped together. The genes within each of these groups are, as a rule, regulated coordinately. On this map six control genes are named individually, as are two sites, *att* (attachment site) and *cos* (cohesive ends).

The region named "replication" includes the two genes required for DNA replication. The region labelled "lysis" includes three genes whose products lyse the bacterium. The "recombination" region contains some ten genes, including two whose products integrate the phage chromosome into the host chromosome during lysogenization and excise it during induction. The approximately ten "*head*" genes encode proteins that construct the phage head and some 12 more the phage tail. We will identify some of these individual genes as we discuss λ growth.

Circularization

The genetic map is shown as a circle because the λ chromosome, a rod in the phage particle, circularizes immediately upon being injected into the bacterium, as shown in Figure 3.2. The ends of the λ chromosome—called cohesive ends—are joined by a bacterial enzyme, producing a pair of continuous intertwined cir-

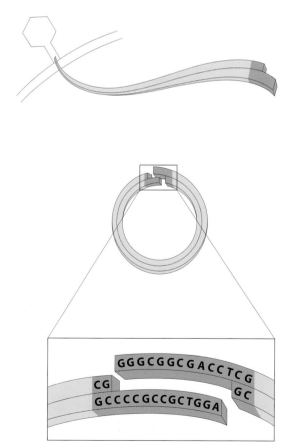

Figure 3.2. Circularization of the λ chromosome. The "sticky ends" of the λ chromosome are 12 bases of single-stranded DNA that emerge, one from each strand, at the ends of the molecule. They pair spontaneously, and bacterial enzymes link the strands together to produce a continuous circular double-stranded DNA molecule.

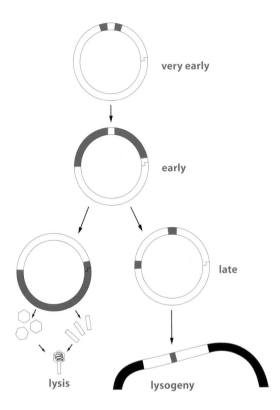

Figure 3.3. Patterns of gene expression. The genes shown in blue are on at each of the indicated stages of growth. Genes of related function are turned on and off together. These coordinately regulated genes lie in contiguous blocks except at the late stages of lysogenic growth where only two genes, *cl* and *int*, are active.

cular DNA strands. The joining brings together the lysis and the tail genes. At the end of lytic growth, when the new phage chromosomes are packaged into new phage heads, the ends of the chromosomes are separated.

Gene Expression

The patterns of gene expression during λ growth are summarized in Figure 3.3. The first two stages of development—very early and early—occur before the decision to lyse or lysogenize is made. We can summarize the pattern of gene expression at each stage as follows:

Very early

Only genes *N* and *cro* are on.

Early

The list of active genes is extended to include the recombination genes and the DNA replication genes.

Late

Here the pathway splits.

- If the phage chromosome is growing lytically, the various early genes are off and the heads, tails, and lysis genes are on. New phage particles are formed and released when the cell lyses.

- If the phage is lysogenizing, only two genes are on—*cI* and *int*. The *int* gene product, located in the recombination region, integrates the phage chromosome into the host chromosome. Finally, in the lysogen, only the repressor gene, *cI*, is active.

Integration

Figure 3.4 shows how a single recombination event integrates the λ chromosome into the much larger host chromosome, a process that occurs only if the phage is lysogenizing the host. The two DNA molecules are each broken once and their ends rejoined to form the single continuous structure in the lysogen. The site of the crossover on the phage DNA is within a region called *attP* (for phage attachment site) and the bacterial site is called *attB*. Upon induction of a lysogen this process is reversed and the λ chromosome circularizes as it is excised from the host chromosome.

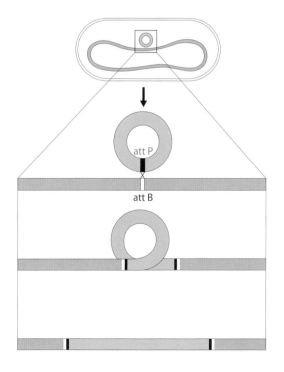

Figure 3.4. Integration. The integration reaction occurs between the phage (*attP*) and bacterial (*attB*) sites to produce two hybrid *att* sites that flank the integrated phage chromosome, the prophage. Excision, the reverse of integration, yields two circular DNA molecules, the phage and bacterial chromosomes, from the circular chromosome of the lysogen. The picture omits the proteins that catalyze the reactions. The Campbell model, which accurately describes the way DNA integrates into and excises from the host chromosome, is named after the man who thought of it.

The integration and excision reactions, shown in the figures as involving only DNA molecules, are driven in the cell by phage proteins that work in conjunction with proteins of the host. The phage protein Int drives the integration reaction, and the combined efforts of Int and Xis promote the excision reaction.

CONTROL OF TRANSCRIPTION

We now describe the activities of lambda regulatory proteins at the various stages of λ growth pictured in Figure 3.3. For each stage a figure shows the active regulatory protein(s) and the relevant mRNAs synthesized. The first two stages are identical whether the phage is to lyse or lysogenize.

Very Early

The host RNA polymerase binds to two promoters on the phage chromosome, P_L and P_R, and begins transcription. Figure 3.5 shows that these transcribing RNA polymerase molecules stop just at the end of N and cro, respectively. The N and cro mRNAs are translated into their respective proteins.

Early

The N protein is a positive regulator. Figure 3.6 shows that it turns on genes to the left of N including cIII and the recombination genes, and genes to the right of cro including cII and the DNA replication genes O and P, and Q. It does not turn on the head, tail, and R genes at the efficiency needed for lytic growth.

The action of N as a positive regulator is entirely different from that of repressor as described in Chapter One. N works by enabling RNA polymerase to transcribe

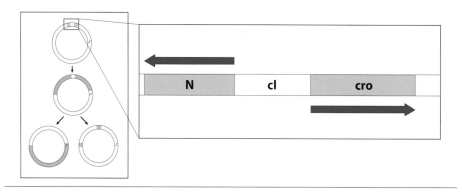

Figure 3.5. Very early events. Very early after infection the *E. coli* RNA polymerase transcribes genes N and cro from different strands of the DNA.

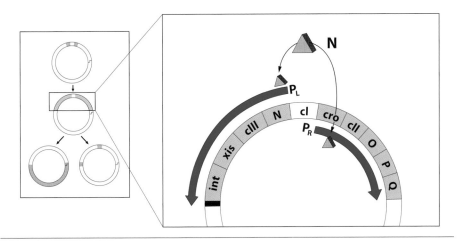

Figure 3.6. Early events. N protein turns on the early genes to the left of *N* and to the right of *cro*. The pyramid representing N protein is shown hovering near the beginning of the leftward mRNA, but further downstream in the case of the rightward mRNA. This is explained in the text.

through regions of DNA that would otherwise cause the mRNA to terminate, and therefore N is called an anti-terminator. In the presence of N, the *N* and *cro* mRNAs are extended, effectively turning on the flanking genes.

How N works is not known in detail. We do know that it recognizes a specific sequence called *Nut* (for *N* utilization); as polymerase passes over this sequence it is evidently modified by N so that it ignores certain (but not all) further termination signals. There is one *Nut* site between P_L and the beginning of *N*, and another just to the right of *cro*. Figure 3.7 emphasizes that the site recognized by N (*Nut*) is distinct from the site where anti-termination actually occurs.

At this point the pathway bifurcates. During lytic growth Q and Cro proteins are active, while for lysogeny CII, CIII, and finally CI proteins are active. We will describe the activities of these two sets of regulatory proteins separately and then return to the question of how the decision is made to express one or the other set.

Late Lytic

Figure 3.8 shows that the Q protein turns on the late genes—those for lysis and for production of heads and tails. Q works like N except that Q anti-terminates specifically a small RNA begun at a promoter called $P_R{'}$, located just to the right of Q. The short (terminated) mRNA is synthesized by the bacterial RNA polymerase. When anti-terminated by Q, this mRNA extends around the circular phage chromosome through the *head* and *tail* genes.

In the meantime Cro first binds to O_R3 as described in Chapter One to prevent further synthesis of repressor from P_{RM}. It then binds to a second operator, O_L, to

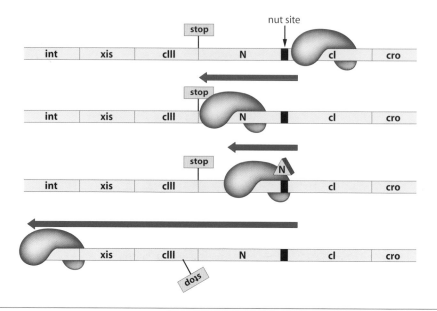

Figure 3.7. The action of N. If no N protein is present polymerase ignores the *Nut* site and falls off the DNA, releasing the mRNA, when it reaches the stop signal. But in the presence of N polymerase becomes a juggernaut as it passes over *Nut* and ignores the stop signal.

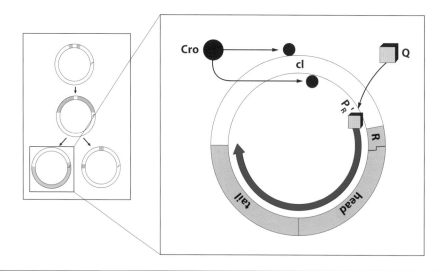

Figure 3.8. Late lytic events. The site Q recognizes, *Qut*, lies very near the beginning of the long transcript that initiates at P_R'. The Q-modified polymerase transcribes the late genes into a single long transcript.

turn off transcription initiated at P_L, and to the remaining sites in O_R to repress P_R. Thus, following a burst of synthesis, these early leftward and rightward transcripts are repressed by Cro. At the late stage, sufficient Q has accumulated to activate production of the protein coats that wrap up the newly synthesized λ DNA molecules. The cell is lysed and a new crop of phage produced.

Late Lysogenic

Figure 3.9 shows that the *cII* gene product turns on *cI* and *int*. The CII protein works like λ repressor in its role as a gene activator—it encourages RNA polymerase to bind and begin transcription at two promoters that would otherwise remain silent: P_{RE} and P_I.

We now can resolve a question that arose in Chapter One. We saw that repressor turns on its own gene, thereby maintaining the production of repressor in a lysogen. How then does repressor synthesis begin in the absence of repressor? The answer is that *cI* can be transcribed from either of two promoters, one of which is activated by repressor, the other by CII protein.

Recall that in a lysogen, repressor stimulates transcription from P_{RM} (promoter for *repressor maintenance*). But in a cell lacking repressor, the CII protein causes polymerase to transcribe the *cI* gene from a different promoter—P_{RE} (promoter for

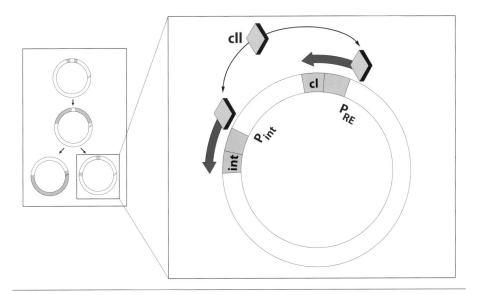

Figure 3.9. Late events in establishing lysogeny. CII protein directs transcription of the two genes needed for finally establishing lysogeny. The early genes are probably still on at this stage but these transcripts have been omitted from the figure.

repressor establishment). Beginning at P_{RE}, polymerase transcribes leftward to the end of the repressor gene *cl*. When this mRNA is translated it produces repressor (but not Cro, because *cro* is transcribed "backwards" here).

Beginning at another promoter, P_I, CII causes polymerase to transcribe leftward the *int* gene, the product of which integrates the phage chromosome into the host chromosome.

Repressor, translated from the mRNA initiated at P_{RE}, binds to O_L and O_R. Repressor bound to these two sites turns on transcription of its own gene from P_{RM} as it turns off all the other phage genes. In a lysogen the late genes are off because there is no Q protein, and the *Q* gene is off because the *N* gene is off.

Lambda's regulatory cascade is controlled by the action of regulatory proteins acting at only a few sites on the chromosome. This is made possible by the fact that genes encoding related functions are grouped together and are transcribed in the same direction.

THE DECISION

Having described the two developmental pathways available to an infecting λ phage, we now must ask: what determines which pathway is taken? What factors drive the system toward lysis or lysogeny?

We do not have a complete understanding of these matters but we can construct a plausible scenario. Briefly put, the decision is effected by a single protein—CII. Figure 3.10 summarizes the situation. If CII is highly active the infecting phage

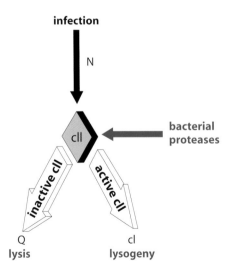

Figure 3.10. The lysis-lysogeny decision. Host proteases regulate the level of activity of CII protein. Although CIII protein is not shown here, the host factors may exert their effects by working on CIII, which protects CII. It is likely that other host proteins regulate translation of the CII mRNA as well.

lysogenizes; otherwise it grows lytically. Once the decision is made, all other steps of λ growth are determined by one or the other of λ's developmental programs.

The activity of CII is determined by environmental factors. As suggested in Figure 3.10, this protein is unstable—bacterial proteases can attack and destroy it. Environmental conditions influence the activities of these proteases. Growth in rich medium, for example, activates the proteases, whereas starvation has the opposite effect and, consequently, λ more frequently lysogenizes starved cells. This makes sense because starving cells are deficient in components necessary for efficient lytic λ growth.

Lambda's CIII protein also helps establish lysogeny—its role is to protect CII from degradation. The protective effect of CIII is not foolproof, and under some environmental conditions CII is largely inactivated even in the presence of CIII. But if CIII is absent, CII is virtually always inactivated and the phage can grow only lytically.

Next, we compare the outcome of infection of cells in which the activity of CII-destroying proteases is high in one case, and low in the other.

- In those cells in which CII is rapidly degraded—protease levels are high—no repressor is synthesized. Q and Cro proteins are synthesized, transcription proceeds as in Figure 3.8, and lytic growth ensues. We shall see that the phage also has a mechanism to diminish the amount of Int protein made from the mRNA that initiates at P_L.

- In those cells in which CII is highly active—low protease levels—transcription of cI and int from the promoters P_{RE} and P_I, respectively, proceeds at a high rate as illustrated in Figure 3.9. Int protein integrates the phage chromosome, and repressor binds to O_L and O_R to turn off all the phage genes except cI. Lysogeny is established.

CONTROL OF INTEGRATION AND EXCISION

The integration reaction requires a single phage protein—Int—but the reverse reaction, excision, requires both Int and Xis. The genes encoding these two proteins lie adjacent on the map, and two different kinds of controls ensure that they are expressed as needed. One involves the action of the lysogeny-promoting protein CII; the other involves modulation of mRNA function.

To set the stage for understanding the logic behind these control mechanisms, let us examine three scenarios: the first is an infecting phage chromosome destined to lysogenize the cell, the second is an infecting phage chromosome destined to multiply lytically, and the third is an integrated prophage about to respond to induction, that is, to be excised and begin lytic growth.

Case 1—Establishing Lysogeny

To ensure integration, the phage should direct synthesis of as much Int and as little Xis as possible. As summarized in Figure 3.11, CII protein turns on transcription of *int*, but not of *xis*. Just as it activates P_{RE} (to turn on repressor synthesis), so, at the same stage of growth, does it work at P_I. The P_I mRNA terminates at the end of the *int* gene, and its translation produces high levels of Int. Some Int and some Xis might be produced from the mRNA extending from P_L, but the action of CII ensures that Int is in excess in the lysogenic pathway.

Case 2—Lytic Growth

In this case, the phage emphasizes production of Xis over that of Int so that the replicating phage chromosomes avoid integrating unnecessarily into the host chromosome. We imagine CII to be relatively inactive along this pathway and we focus our attention on the P_L transcript which covers both the *int* and *xis* genes. Here, for the first time, we encounter regulation, not at the level of initiation of transcription of DNA to RNA, but rather at the level of mRNA stability.

Recall that the anti-terminator N causes mRNA originating at P_L to extend beyond the *N* gene into the *int* and *xis* genes. N, in fact, has turned the RNA polymerase into a juggernaut that ignores the signal at the end of *int* that would otherwise stop an RNA polymerase. Thus N-modified polymerase continues onward and copies a region called *sib*, as shown in Figures 3.11 and 3.12.

Now a remarkable fate is in store for the *sib*-bearing mRNA: it is sequentially destroyed, beginning at a position near *sib* and proceeding back toward N. Because information encoding *int* is destroyed before that encoding *xis*, this mRNA produces more Xis than Int. This control is called retroregulation.

Case 3—Induction

In the third case, induction of a lysogen, the phage must make *both* Int and Xis to ensure its escape from the chromosome following induction. But because we are

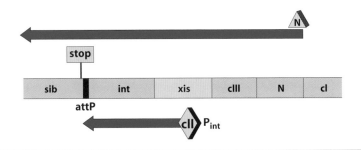

Figure 3.11. CII-stimulated transcription of *int*. The promoter for the *int* gene, P_{int}, lies within the *xis* gene. Therefore Int but not Xis production is stimulated by CII.

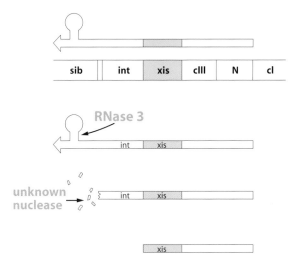

| sib | int | xis | cIII | N | cI |

Figure 3.12. Retroregulation. The mRNA copy of *sib* forms a hairpin that attracts the bacterial enzyme RNase III, which cleaves the hairpin. Other bacterial RNase molecules then chew the mRNA, beginning at the cleavage site.

witnessing the onset of lytic growth we cannot depend on CII to provide the requisite Int. The solution is as follows: transcription from P_L once again produces a mRNA encoding both Int and Xis, but in this case the *sib* region is *not* copied onto the end of the mRNA. As can be seen in Figure 3.13, the process of integration of the phage chromosome separates *sib* from *int* and so the mRNA originating at P_L bears no *sib* sequence. Consequently, both *int* and *xis* are translated from this mRNA.

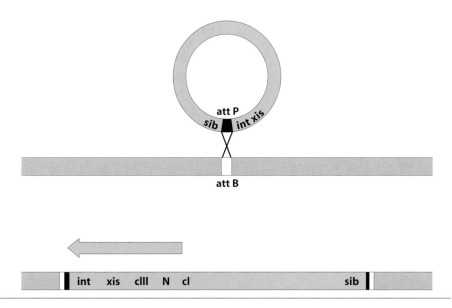

Figure 3.13. Integration and gene order. Integration (recombination at *att*) has separated sib from *int*.

OTHER PHAGES

Lambda is a member of a group of phages that grow in and lysogenize *E. coli*. For some of these phages the sequence of the DNA bears no obvious relation to that of λ. Nevertheless, the general organization of their genes and the mechanisms for controlling them is very similar to λ.

Thus the map of Figure 3.1 applies with slight modification to a number of different phages. Each one of these phages has, for example, a different sequence at the *att* site and therefore attaches to a different site on the bacterial chromosome. In each case the *int* and *xis* gene products are designed to catalyze a particular integration-excision reaction and none other. To take another example, a phage other than λ makes a Q-like protein that anti-terminates the small RNA made from its chromosome, but does not anti-terminate λ's small RNA. To take yet another example, each of these phages elaborates its own repressor and Cro that act on its own operators, but not on those of λ.

Whether we consider overall organization, as in this chapter, or molecular details, as in Chapters One and Two, it is apparent that the logic of the control mechanisms is preserved in the face of diversity of specific sequences. The various phages evolve together, occasionally reassorting parts by genetic recombination. For example, if we replace the λ *att-int-xis* region with the corresponding DNA of another phage, we produce a phage identical with λ except that it integrates at a different point on the host chromosome.

This reshuffling of cassettes of information requires that in general the DNA bases encoding the sites of action of regulatory proteins lie close to the genes encoding the regulatory proteins. Thus repressor and Cro bind to the operators that flank the *cI* gene; O and P work on the origin (*ori*) located in the *O* gene itself; *int* and *xis* work on att; Q works on a site (*Qut*) near *Q*; and the N recognition sites (*Nut*) lie near *N*.

THE SOS RESPONSE

Induction of λ is only a part of a larger set of responses that bacteria mount to ultraviolet irradiation. An array of 10–20 bacterial genes is turned on; at least some of which help the bacterium survive the radiation. This multiple gene activation is called the SOS response.

The mechanism by which ultraviolet light induces these bacterial genes is the same as that involved in the induction of phage growth in a lysogen. Recall from Chapter One that ultraviolet light damages DNA. This insult prompts RecA, a protein ordinarily involved in facilitating recombination between DNA molecules, to become a special kind of protease: it cleaves λ repressor, thereby triggering induction as shown in Figure 1.21. RecA cleaves at least one other repressor as well, a host cell protein named LexA.

LexA is similar to λ repressor in both structure and function. The LexA monomer is folded into two domains, an amino domain which recognizes DNA and a carboxyl domain that holds dimers together. LexA represses a set of genes whose products help the bacterium survive irradiation, as shown in Figure 3.14. Ultraviolet irradiation turns these genes on by inactivating LexA through the action of RecA. RecA cleaves LexA, just as it cleaves λ repressor, at a specific bond in the stretch of amino acids connecting the two domains.

Lambda's repressor, like that of other phages that lysogenize *E. coli*, exploits the cell's SOS response. The repressed phage senses that its host has been damaged and it quickly begins lytic growth.

Ultraviolet light is not the only agent that causes induction of SOS genes. In general, any carcinogen, suitably activated, will work as well. Benzpyrene is an example of a compound that causes tumor formation in animals. Benzpyrene is itself inert, but when modified by enzymes found in an animal's liver, it reacts with and damages DNA. Benzpyrene added directly to bacteria has no effect, but if it is first modified by the liver enzymes it efficiently induces the SOS response.

The ability of an agent to induce SOS is correlated with its ability to cause tumor formation in animals, and although there is some debate concerning the reliability of this correlation, bacteria are used as sensitive detectors of suspected carcinogens. Two convenient assays are the induction of phage growth in lysogens and the production of mutants among non-lysogens. The latter assay is the basis of the well-known Ames Test for carcinogens. In both assays the primary event, following DNA damage, is the induction of gene expression: phage induction requires that RecA activate phage genes by cleaving the phage repressor, and mutagenesis requires that RecA activate bacterial genes by cleaving LexA.

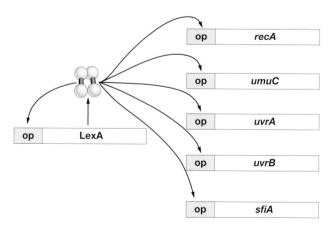

Figure 3.14. The action of LexA and the SOS response. Some of the genes repressed by LexA are shown here. *uvrA* and *uvrB* encode proteins that repair ultraviolet-damaged DNA. The umuC protein is required for formation of mutant bacteria in response to ultraviolet light. Op stands for the LexA operator, which is found adjacent to the promoters of each of these genes.

λ PATHWAYS AND CELL DEVELOPMENT

We have considered the alternate forms of λ gene expression as a miniature problem in development. Here we draw three parallels between these processes and the development of a eukaryotic organism from a fertilized egg. This exercise helps us to summarize and recast some of the essential features of gene regulation in λ.

Regulatory Genes

The study of higher organisms has revealed the existence of genes whose role is to control the expression of other genes. We see a particularly striking example in the fruit fly *Drosophila melanogaster*. This organism develops from a fertilized egg in three stages, first forming the embryo, then the larva, and finally the adult. Both the larva and adult are composed of some 15 segments, which together form the head, thorax, and abdomen. Each segment, especially if it is part of the thorax or abdomen, is easily distinguished from the other segments in both the larva and the adult.

   To illustrate how differentiation of the segments is controlled we focus on two adjacent thoracic segments of the larva, T2 and T3. One particular gene product, encoded by a gene called *Ubx*, is expressed in T3 but not in T2. If this function is missing from T3, T3 develops identically to T2; if it is present in both segments, T2 develops identically to T3. Thus addition of this *Ubx* product to a developmental pathway that would otherwise produce segment T2 causes formation of segment of T3.

   Homeotic genes, of which *Ubx* is an example, direct groups of cells to differentiate in a particular way. We suspect that, early in development, a particular set of homeotic genes is turned on in each segment. These genes remain on through adult life and maintain particular aspects of the pattern of gene expression characteristic of that segment. There is no proof that each homeotic protein is a direct regulator of genes but many workers now believe this is a reasonable idea.

   Returning to λ, we might say that *cl* is analogous to a homeotic gene. This notion is illustrated by consideration of a hypothetical ancestral λ phage missing the *cl* gene. The phage would grow well, but only in the lytic mode—in other words, it would realize only one program of gene expression. Addition of gene *cl* (along with *cll* and *clll*) would allow the phage to achieve a second state of gene expression. We recall that *cl* is turned on during the early phase of the establishment of lysogeny by the action of *cll*, abetted by *clll*. Once on, expression of *cl* is self-perpetuating and it, in turn, maintains the pattern of gene expression characteristic of a lysogen. Gene *Q* is another homeotic-like function of λ because it, like *cl*, is required for only one of the alternate modes of λ growth.

   Not only are genes *cl* and *Q* required for one or the other developmental pathway, both their products in large part determine the pathway. Thus if *Q* protein were introduced into a lysogen, the late genes would be turned on and lysis would ensue. If repressor were added to a phage beginning its lytic cycle, growth would be inhibited. In contrast to these genes, λ's *N* is required both for lytic growth of the phage and for establishment of lysogeny. Because the phage cannot proceed down either developmental pathway in the absence of N, we would not call *N* a homeotic gene.

In thinking about development it is sometimes useful to distinguish between "commitment" and overt "differentiation" of cells. This distinction reflects the observation that cells appearing by casual inspection to be identical early in development nevertheless later differentiate, under apparently identical conditions, to form distinct cell types. They are therefore said to be committed to a particular pathway of differentiation when the appropriate stimulus is later received.

One form of commitment may be the activation of certain genes, perhaps regulatory genes. The pattern of activation could determine the subsequent response to an external signal. Overt differentiation would then result from a further change in the pattern of gene expression.

By analogy, a λ-lysogen is indistinguishable from an uninfected bacterium by casual inspection, but the former and not the latter is "committed" to lyse and release phage when it encounters an inducing signal in the environment. The lysogen, of course, has genes that are lacking in the non-lysogen, but it is not difficult to imagine analogous forms of commitment involving only differential gene expression. To take a simple example, the regulatory proteins of a cell might turn on a gene encoding a hormone receptor. That cell, but not its neighbor bearing a quiescent hormone receptor gene, would respond to the hormone, perhaps to undergo a further round of changes in the pattern of gene expression.

Switches

At various stages during the development of a higher organism, environmental signals trigger one or another path of gene expression. To illustrate one example we briefly inspect a *Drosophila* embryo at the early blastula stage. The original egg has not yet divided but the nucleus has, and inside the egg there are 4,000 nuclei arrayed along the inside surface. In a short time, these nuclei will become committed to form cells constituting specific parts of the larva.

How are these early nuclei instructed to turn on the appropriate sets of genes that commit them to the proper pathways of differentiation? It is believed that the egg contains asymmetric distributions—perhaps gradients—of molecules that trigger gene expression. Thus the anterior part of the egg presents a different environment to the nuclei than does the posterior part. Similarly, it is suspected that the ventral part of the egg differs from the dorsal part in the concentration of some substance(s). Each cell nucleus must decipher its *position* in the egg by sensing the concentrations of these factors, thereby triggering a specific pathway of gene expression.

Similarly, at later stages of fly development it is believed that cells make further genetic decisions based on their positions in the developing embryo. In these cases, it is imagined, cells sense their positions according to the concentrations of factors produced by other cells, and turn on or off the appropriate genes.

Turning to λ, we recall that there are two stages in λ's life cycle at which extra-cellular signals modify gene expression. In both cases the important effect is on gene *cI*, but the signal is sensed and transmitted by different proteins in the two cases. In the first case, upon induction of a lysogen by ultraviolet irradiation, the protein RecA inactivates repressor. In the second case, growth conditions at the time of infection modify the activity of the CII protein, which must turn on transcription of *cI* for the establishment of lysogeny.

In both cases of environmental influence, λ uses extracellular signals to decide whether to follow one or another mode of growth. We might say that in each case a biphasic switch is thrown into one or the other state by extracellular conditions. Lambda's induction switch, the one we understand best, is designed so that rather modest differences in the level of repressor determine whether a lysogen induces. It is reasonable, in other words, to imagine that a factor of only 5 or 10 in the distribution of a regulatory protein would flip a *Drosophila* genetic switch if it were designed like a λ switch. A switch of this design virtually never flips accidentally. Thus, if we were to follow a single line of descendants from a λ-lysogen, looking at only one daughter at each cell division, we would have to wait 5–10 years before seeing a cell lyse in the absence of an inducing signal.

Patterns of Gene Expression

We imagine that during development a higher organism expresses genes in many different combinations. A given gene, for example, might be turned on at different stages of development and in different tissues, in each case as a member of a different set of activated genes. This requires the imposition of multiple controls on single genes.

The alcohol dehydrogenase (ADH) gene of *Drosophila*, for example, is expressed in one set of tissues in the larva and in a different, partially overlapping set in the adult. In this case we have the beginnings of an understanding of the mechanisms involved. Thus, two promoters direct transcription of this gene, and these promoters are under separate developmental control. Both promoters are active in the embryo, but only one is active in the larva, and only the other in the adult.

In our study of λ we have seen that several of its genes are expressed as members of one or another set of genes according to the phase of growth. Gene *cI*, for example, is expressed concurrently with *int* as lysogeny is being established, but in an established lysogen *cI* is on and *int* is off. Gene *cI* is transcribed from different promoters in these two conditions, from a CII-stimulated promoter as lysogeny is being established, and from a repressor-stimulated promoter in a lysogen.

Considering the adjacent genes *int* and *xis*, we recall that there are three patterns of expression—during lysogenization predominantly *int* is expressed, during lytic growth predominantly *xis*, and upon induction *int* and *xis* are expressed coordinately. These three scenarios are realized using two promoters to transcribe *int* and imposing "retroregulation" on the *int* mRNA as well.

The following example, taken from bacteria, shows an important effect that can be achieved by superimposing two forms of control on a single gene. That is, in this case, the same compound activates one or another set of genes depending on the conditions.

The lactose (*lac*) genes and galactose (*gal*) genes code for enzymes that are responsible for metabolizing the sugars lactose and galactose, respectively. Each set of genes is regulated negatively by its own repressor. The presence of lactose in the medium releases the *lac* repressor from the DNA, and similarly, the *gal* repressor is inactivated by galactose in the medium.

These genes, together with many other genes in *E. coli* are activated by CAP, a positive regulator protein. CAP binds to DNA and activates transcription only in the presence of cyclic AMP, and only if the corresponding repressor is released from the DNA. Thus the small compound cyclic AMP activates the *lac* genes if lactose is present, but activates the *gal* genes if galactose is present.

The regulatory effect of cyclic AMP is responsible also for another regulatory effect. The level of cyclic AMP is controlled in turn by glucose. If glucose is present, cyclic AMP synthesis is depressed, and so neither the *lac* nor the *gal* genes can be activated. The result is that, when offered a choice of sugar sources, the bacterium metabolizes glucose in preference to either lactose or galactose.

FURTHER READING: RELATED REVIEWS

All of the following reviews are found in the book Lambda II, *R.W. Hendrix, J.W. Robert, F.W. Stahl, and R. Weisberg, eds. (New York: Cold Spring Harbor).*

1. Campbell, A. and Botstein, D. (1983). Evolution of the lambdoid phages, pp. 365–381.

2. Echols, H. (1983). Control of integration and excision, pp. 75–93.

3. Friedman, D. and Gottesman, M. (1983). Lytic mode of λ development, pp. 21–53.

4. Hershey, A.D., and Dove, W. (1983). Introduction to λ, pp. 3–13.

5. Weisberg, R. and Landy, A. (1983). Site-specific recombination in phage λ, pp. 211–251.

6. Wulff, D. and Rosenberg, M. (1983). Establishment of repressor synthesis, pp. 53–75.

The following reviews discuss various aspects of gene regulation during development in Drosophila.

7. Garcia-Bellido, A., Lawrence, P.A., and Morata, G. (1979). Compartments in animal development. *Sci. Amer.* 241, 102–110.

8. Gehring, W.J. (1985). The molecular basis of development. *Sci. Amer.* 253, 153–162.

9. Kauffman, S., Shymko, R., and Trabert, K. (1978). Control of sequential compartment formation in *Drosophila. Science* 199, 259–270.

10. Lewis, E.B. (1982). Control of body segment differentiation in Drosophila by the bithorax gene complex. In *Embryonic Development: Genes and Cells*, ed. M. Burger. (New York: Liss), pp. 269–288.

CHAPTER FOUR

HOW DO WE KNOW— THE KEY EXPERIMENTS

The first few sections of this chapter describe some of the major experiments that provide important background for our story. In sections that follow, many of the figures of Chapter One and Two are keyed to selected experiments. The detailed experimental data will be found in the references listed at the end of this chapter and in other papers cited in those papers. This chapter does not deal with the experimental bases of the descriptions of Chapter Three. For that the reader must explore the general references cited in that chapter.

THE REPRESSOR IDEA

The idea that λ encodes a repressor, a molecule that turns off the other phage genes, was formulated by the French scientists F. Jacob and J. Monod and their colleagues in the late 1950s and early 1960s. The argument was based primarily on genetic studies, beginning with a simple observation concerning growth of the phage. This section describes three such sets of observations and their explanations according to the repressor hypothesis.

Clear and Virulent Mutants

Observations

One way to observe λ phage growth is to mix a few phages with a million or so bacteria and pour the mixture on an agar plate. The bacteria grow, forming a lawn across the surface of the agar. But as the bacteria grow so do the phages, and each viral particle gives rise to one plaque, a hole, in the bacterial lawn, as illustrated in Figure 4.1.

The plaques formed by wild-type λ phages are turbid (cloudy) because although many of the infected bacteria lyse as the plaque expands, some do not, and instead overgrow the plaque. When the bacteria from the center of the plaque

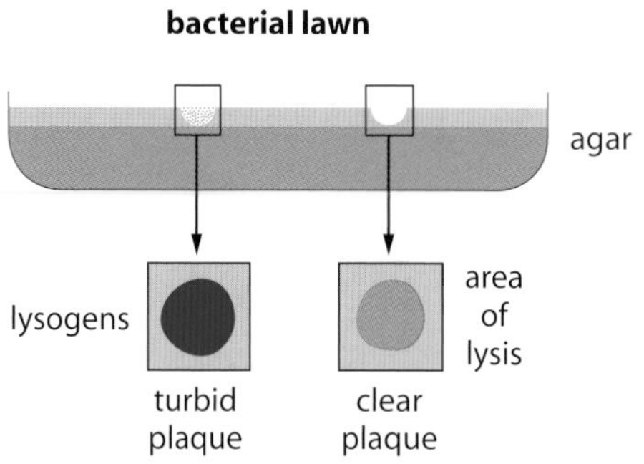

bacterial lawn

Figure 4.1. Plaques formed by growth of λ. Lambda phage are mixed with bacteria and poured on an agar plate. Each infected cell lyses and releases progeny phage that infect nearby bacteria, a process that is repeated many times. Lysogenic (and hence immune) cells grow in the center of a plaque formed by wild-type λ, making the plaque turbid.

are picked and regrown they are found to be immune to infection by other λ phages. These immune cells are lysogens—when irradiated with ultraviolet light they lyse and produce a crop of new phages.

But occasionally one observes a clear (not cloudy) plaque. Each clear plaque arises from a mutant λ that is deficient in its ability to lysogenize, but which otherwise grows normally. The clear plaque is an area of pure lysis and contains no lysogenic bacteria.

Most of these clear mutants bear a lesion in one of three separate genes: *cI*, *cII*, or *cIII*. It is possible to isolate a rare lysogen of a *cII* or *cIII* mutant; once isolated the lysogen is stable, immune, and inducible by ultraviolet light. But it is never possible to isolate a lysogen of a *cI* mutant.

Lambda *cI* mutants cannot establish lysogeny but they, like wild-type and like the other clear mutants, are unable to lyse λ-lysogens. A different kind of λ called λ*virulent* (λvir) grows on lysogenic as well as non-lysogenic bacteria. These mutants, which form clear plaques on lysogens and non-lysogens alike, arise in the population much more rarely than do the usual λ*clear* mutants. Their rarity, compared with λ*clear* mutants, suggested that λvir contains more than one mutation.

Explanation

In a lysogen, the product of gene *cI*—the repressor—turns off all the phage genes except *cI*, including those required for lytic phage growth. Genes *cII* and *cIII* function early in infection to establish lysogeny but are not required for its maintenance. We now know that CII, abetted by CIII, turns on transcription of *cI*. λvir bears mutations in the sites of action of repressor, the operators; repressor cannot bind to these mutated operators, and so a λvir grows lytically even in the presence of repressor. For a λ phage to become virulent it must acquire mutations in two operators, O_L and O_R.

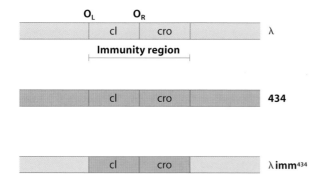

Figure 4.2. The immunity region. The λ- and 434-immunity regions comprise only a few percent of the length of each phage chromosome.

Immunity and Heteroimmunity

Observations

Lambda is but one of a large group of phages that grow on and lysogenize *E. coli*. Phage 434 is a typical example of a λ-like phage. It forms turbid plaques on *E. coli*, and clear and virulent mutants arise as in the case of λ. But a striking characteristic of 434-lysogens is that, although they are immune to 434 phages, they are *not* immune to λ phages; in fact, λ growth on a 434-lysogen is indistinguishable from growth on a non-lysogen. Similarly, λ-lysogens are sensitive to 434 phages. We say that λ and 434 are heteroimmune.

Genetic recombination experiments showed that the immunity characteristics of each phage are determined by a short piece of the phage chromosome called the immunity region. Figure 4.2 shows that it includes the *cI* and *cro* genes and the sites on which their products act.

A hybrid chromosome bearing all the genes of phage λ except the λ-immunity region, which has been replaced with that of 434, has the immunity characteristics of phage 434. Thus λimm^{434} or 434hy as it is variously called, forms lysogens that are immune to 434 phages but not to λ phages, and the hybrid itself can grow on λ-lysogens but not on 434-lysogens. Table 4.1 summarizes the growth of various phages on various lysogens.

Table 4.1. Plaques formed by various phages when mixed with each of three bacterial strains and grown as in Figure 4.1.

	λ	λ*cI*	λ*vir*	434	λimm^{434}
E.coli K	tu	cl	cl	tu	tu
K(λ)	–	–	cl	tu	tu
K(λimm^{434})	tu	cl	cl	–	–

"cl" means the plaque is clear, "tu" means the plaque is turbid, and "–" means no plaques are observed. K (λ) is the designation for a λ-lysogen of *E. coli* strain K. The entry using K(434) would be identical with that given for K(λimm^{434}).

Explanation

Phage 434 encodes a repressor with a specificity different from that of λ, one that binds 434 operators but not λ operators. Conversely, λ repressor cannot recognize 434 operators. The immunity region defined by the λ-434 experiments includes *cl* and, we now know, *cro*, as well as the left and right operators and promoters. The fact that the immunity region does not include genes *cll* or *clll* shows that λ's Cll and Clll can turn on 434's *cl* gene, and vice versa, 434's Cll and Clll can turn on λ's *cl* gene.

Asymmetry in Bacterial Mating

Observations

Lambda-lysogens, as well as lysogens of other phages, behave asymmetrically in bacterial mating experiments. In a bacterial mating, male *E. coli* attach to and then inject their DNA into female cells. If the female is lysogenic, or if both parents are lysogenic, the conjugation proceeds unimpeded, and the immediate product is a zygote consisting of a female cell bearing some additional male genes. But if the male is lysogenic and the female is not, or if the female is lysogenic for a phage of a different immunity type, many of the females lyse, releasing phage. Figure 4.3 illustrates this process of zygotic induction.

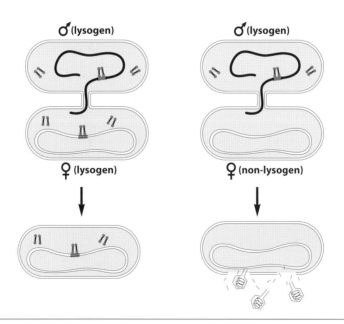

Figure 4.3. Zygotic induction. In a bacterial mating the male genes enter the female in a defined order. Different male strains begin the injection starting from different positions on the circular *E. coli* chromosome. Proteins, including λ repressor, are not transferred to the female during mating.

Explanation

Figure 4.3 shows that if the male is lysogenic but the female is not, the cytoplasm of the former but not that of the latter, contains the repressor. Only DNA, not cytoplasm, is transferred to the female. The phage chromosome, integrated into that of the host, is transferred to the female where, in the repressor-free environment, the phage genes are activated.

References

Original experiments on virulence, immunity, and zygotic induction were reported in References 3, 7, 34, and 35, cited at the end of this chapter.

THE REPRESSOR PROBLEM IN THE EARLY 1960s

The early λ experiments were paralleled by an equally important series of experiments analyzing control of synthesis of the bacterial enzymes that metabolize the sugar lactose. Figure 4.4 illustrates the idea that the product of the *lacI* gene encodes a repressor (the *lac* repressor) that recognizes the *lac* operator and turns off several genes. One of these genes, *lacZ*, encodes β-galactosidase, an enzyme that cleaves the sugar. Lactose, or a metabolic derivative thereof, was imagined to bind to and inactivate the *lac* repressor and thereby turn on the *lac* genes.

In the early 1960s, the fundamental tenets of the repressor idea were open to challenge. It could be (and was) argued, for example, that the products of the genes *cI* and *lacI* were not repressors as imagined, but rather were enzymes that catalyze

lac genes

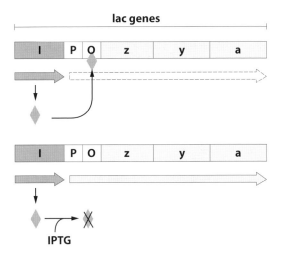

Figure 4.4. The *lac* genes. The *lac* repressor turns off three contiguous genes involved in metabolism of the sugar lactose. The compound IPTG (isopropyl thio β-galactoside) resembles lactose, and it induces these three genes coordinately. Binding of inducer does not destroy repressor but modifies it so that it loses affinity for the operator. The *lac* repressor works only as a negative regulator and its gene, lacI, is transcribed continuously at a low level. The figure omits the CAP protein that, in the presence of cyclic AMP, helps polymerase bind and begin transcription when the lac promoter is available.

other reactions that lead to formation of the "real" repressor. Genetic experiments alone could not eliminate this argument, nor could they eliminate debates concerning the molecular nature of repressors. An early suggestion concerning how repressors might work was that they would bind directly to operator sites on DNA to turn off gene transcription. But other ideas, equally plausible, were also suggested. The operator might have been transcribed as part of the mRNA for example, and then, according to this idea, repressor would bind to the mRNA to turn off translation.

It became clear that it was necessary to isolate the products of regulatory genes. The hope was that these gene products would display some striking property in vitro that would reveal their modes of action. The isolation of the lac and λ repressor was accomplished in 1966 and 1967, and the λ experiments are briefly recounted here to make the point that the combination of genetics and biochemistry can be powerful.

References: 3, 8.

REPRESSOR ISOLATION AND DNA BINDING

The method first used to isolate λ repressor was dictated by the following considerations: repressor is present in very small amounts (a few hundred molecules per cell compared with tens of thousands of molecules of a typical enzyme); there was no known assay for its actitivity; and its molecular composition was unknown other than that it was at least partly protein.

The solution, depicted in Figure 4.5, was to infect two batches of *E. coli* cells with two λ strains that differed only in that one could make repressor (cI$^+$) and the other could not (cI$^-$). The cells were λ-lysogens, and therefore contained λ repressor that would turn off all the genes of the infecting phages except the *cI* gene. In addition the cells were irradiated, before infection, with massive doses of ultraviolet light, a treatment designed to damage the host DNA so that it could not synthesize its own proteins. (The repressor present in the irradiated cells was a mutant form that, unlike wild type, is not inactivated by ultraviolet irradiation.)

In principle, then, one batch of infected cells should synthesize one new protein—λ repressor—but the other batch, infected with the cI$^-$ mutant, should not. Indeed, when one batch of infected cells was labeled with ^{3}H-leucine and the other with ^{14}C-leucine, the cells mixed, opened, and their contents examined, one protein was detected that was labeled with ^{3}H but not ^{14}C. The repressor was only a small fraction of the total protein in the final mixture, but it could be followed in experiments by virtue of its unique radioactivity. But what could this protein—the product of the *cI* gene—do?

A single experiment showed that the cI product bound specifically to a short piece of double-stranded λ DNA known to contain the λ operators. Figure 4.6

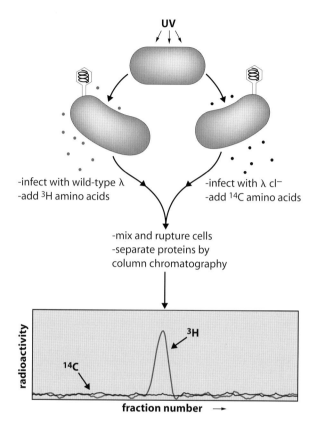

UV

-infect with wild-type λ
-add ³H amino acids

-infect with λ cl⁻
-add ¹⁴C amino acids

-mix and rupture cells
-separate proteins by
 column chromatography

radioactivity

³H

¹⁴C

fraction number →

Figure 4.5. Isolation of λ repressor. The original isolation of λ repressor was really a selective labeling of repressor present in very small amounts in the cells. The rate of repressor synthesis, relative to that of other proteins in the cell, was raised using ultraviolet light as described in the text. The cl⁻ mutant was used to confirm that the labeled protein really was the product of the *cl* gene.

shows how the radioactively labeled *cl* product was mixed with λ DNA in one tube and with λ*imm*⁴³⁴ DNA in another, and the mixtures were sedimented by centrifugation through separate sucrose gradients. In these gradients, λ-sized DNA molecules move from the top toward the bottom of the centrifuge tube much faster than the free repressor, and the two components are easily separated.

The labeled protein was found to travel with the fast moving λ DNA but *not* with the equally fast moving λ*imm*⁴³⁴ DNA. Recall that λ*imm*⁴³⁴ DNA differs from that of λ only in a short region of its chromosome bearing genes *cl*, *cro*, and the operators (see Figure 4.2). The result of the experiment argued strongly that *cl* encodes a protein—the repressor—that binds directly and specifically to the operator. Furthermore, if the DNA were denatured—that is, if the two strands of the helix were separated by heating—no co-sedimentation of protein and DNA was observed. Thus the operator must be double-stranded DNA.

References: 53, 54.

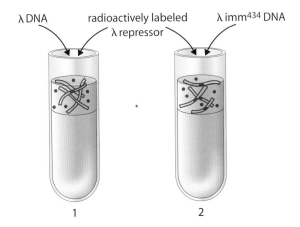

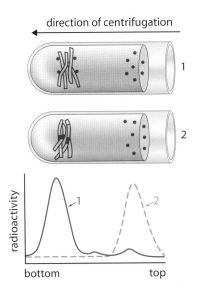

Figure 4.6. Specific binding of λ repressor to λ DNA. The experiment shows that λ repressor binds to λ DNA but not to λ*imm*^{434} DNA. A similar experiment showed that 434 repressor binds λ*imm*^{434} DNA but not λ DNA. Since the two DNAs differ only in their immunity regions, the repressors must be binding to sites within that region. An additional experiment showed that DNA isolated from a λ*vir* phage binds λ repressor much less well than does wild-type λ DNA, confirming that the repressor binds to the genetically defined λ operators.

MAKING MORE REPRESSOR

To study the structure of repressor it was first necessary to construct strains of *E. coli* that make large amounts of repressor so that it could readily be purified. Recombinant DNA technology made this possible. The *cI* gene was cloned—that is, as outlined in Figure 4.7, the gene was excised from the λ chromosome with DNA-cutting enzymes (restriction enzymes) and inserted into a plasmid. (A plasmid is a

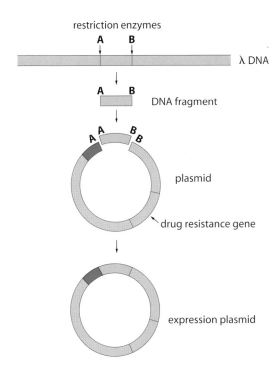

restriction enzymes
A   B
λ DNA

A   B
DNA fragment

A   A   B
A   B   B

plasmid

drug resistance gene

expression plasmid

Figure 4.7. A plasmid that directs syn-thesis of large amounts of repressor. The plasmid bearing the λ *cl* gene abutted to the lac promoter/operator (Op) rather infrequently enters *E. coli* when the cells and the plasmid are mixed. The drug resistance gene encodes a protein that makes the cell resistant to an antibiotic, and so in the presence of the antibiotic only cells containing the plasmid grow. Thus, rare cells bearing the plasmid are easily selected.

circular DNA molecule, much smaller than the *E. coli* chromosome that, in the case represented here, is present in 25–50 copies in each cell. Plasmids replicate independently of the host DNA molecule, and they can be extracted from cells, modified, and reintroduced into other cells.)

But simply increasing the number of copies of the *cl* gene does not greatly increase the amount of repressor synthesized. (Recall from Chapter One that at moderately high concentrations repressor turns off its own gene [by binding to O_R3].) To make very large amounts of repressor, therefore, we needed to replace the ordinary repressor promoter (P_{RM}) with a different promoter, one not subject to control by λ repressor.

As shown in Figure 4.7, we chose the promoter that ordinarily drives the *lacZ* gene. A DNA fragment bearing this promoter as well as the signals required for effi-cient mRNA translation was isolated and abutted to the very beginning of the *cl* gene. The construct, which was introduced into cells as part of a plasmid, directed the synthesis of large amounts of repressor. Current strains, which use a modified form of the *lac* promoter (called *tac*), synthesize as much as 20% of their protein as repressor. The repressor is easily purified from these strains.

References: 12, 15, 56.

THE CLAIMS OF CHAPTERS ONE AND TWO

The repressor is composed of two globular domains held together by a linker of some 40 amino acids (Figure 1.6).

Treatment of the purified repressor with a protease (for example, papain), produces two relatively stable protein fragments as shown in Figure 4.8. Amino acid sequence analysis reveals that one of these fragments is from the amino end of repressor (residues 1–92) and the other from the carboxyl end (residues 132–236).

Both fragments behave as typical globular proteins, and they are easily separated. Tightly folded regions of proteins are generally more resistant to protease than are the more extended conformations, so these experiments indicate that the amino and carboxyl portions of repressor fold into separate domains.

Proteins—or domains of proteins—unfold (denature) at elevated temperatures, each domain unfolding at a characteristic temperature. When unfolding is measured for the separated domains, the amino-terminal domain of repressor denatures at a temperature about 20°C below that at which the carboxyl fragment denatures, as shown in the experiment of Figure 4.9.

When the intact repressor is heated in this experiment, two separate peaks of denaturation are seen that correspond fairly closely to the peaks observed with the separated fragments. Thus, even when held together in the native molecule, the domains denature relatively independently. Nuclear magnetic resonance (NMR) studies also indicate that the amino domain is flexibly attached to the carboxyl domain in the intact repressor.

References: 50, 64.

The repressor dimerizes, largely through interaction between its carboxyl domains (Figure 1.7).

Gel filtration and sedimentation analyses show that at low concentration (10^{-9}M) repressor is predominantly monomer, at higher concentration (10^{-7}M) predominantly dimer, and at very high concentration (10^{-5}M) predominantly tetramer. The separated carboxyl domains (residues 132–236) dimerize and tetramerize about as efficiently as do the intact monomers, but the separated amino domains do not (Figure 4.10). Some interaction between amino domains is evident from the X-ray structure of the amino domain.

The dimerization constant of intact repressor, $K_D = 2 \times 10^{-8}$M (see below), taken with the concentration of repressor in a lysogen (about 4×10^{-7}M) implies that about 95% of repressor in a lysogen is dimer, 5% monomer. The concentration at which one half of the repressor forms tetramers is at least 100-fold higher than the concentration actually found in a lysogen, and so very little repressor tetramer is formed in vivo.

References: 1, 4, 50.

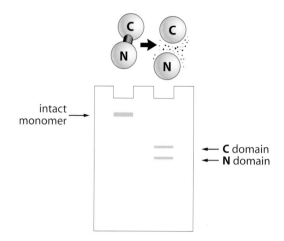

intact
monomer →

← **C** domain
← **N** domain

Figure 4.8. Protease cleavage of λ repressor. One way to define domains in a protein is to see whether parts of it are relatively resistant to digestion by a protease. Papain easily digests residues 93–131 of λ repressor, but the amino and carboxyl regions survive. The domains are separated by gel electrophoresis; they are different sizes and therefore form separate bands on the gel.

Figure 4.9. Denaturation of λ repressor and of the separated domains of λ repressor. Lambda repressor unfolds in discrete steps at two temperatures. These temperatures correspond to those at which the separated domains denature. (Denaturation is measured in a machine called a scanning calorimeter. As proteins denature they absorb heat, and the scanning calorimeter measures the heat absorbed as the temperature is increased.)

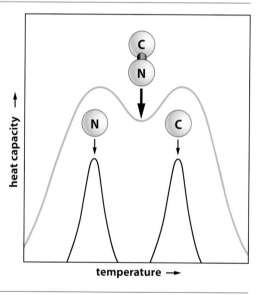

heat capacity →

temperature →

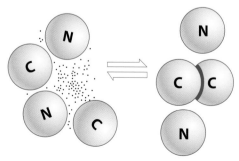

Figure 4.10. Dimerization of cleaved λ repressor. Papain cleavage has little or no effect on the ability of the carboxyl domain of λ repressor to dimerize.

A repressor dimer binds, through its amino domains, to a 17 base pair operator site (Figure 1.9).

Three conclusions are implied by this figure.

- A single operator site binds one dimer of repressor.

- The dimer forms before DNA binding (as opposed to having each monomer attach separately).

- Only the amino domains contact the DNA.

We consider these separately.

A single operator site binds one dimer of repressor

The nitrocellulose filter assay shown in Figure 4.11 is an important tool for investigating DNA-protein binding reactions, providing accurate measures of equilibrium and kinetic parameters.

The assay exploits the fact that double-stranded DNA passes through the filter, whereas repressor (like most other proteins) sticks to the filter. DNA bound to the repressor is trapped on the filter, and if the DNA is labeled with a radioactive isotope (for example, with ^{32}P) it is easy to measure the fraction of DNA bound to the filter (and hence to repressor) under a variety of conditions. For example, at high

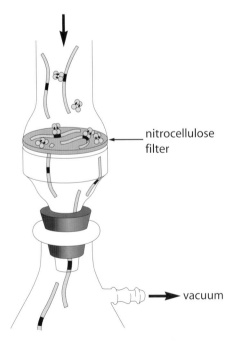

nitrocellulose filter

vacuum

Figure 4.11. The filter-binding assay. In this assay the DNA is radioactively labeled but the protein is not. At higher protein concentration more DNA would stick to the filter.

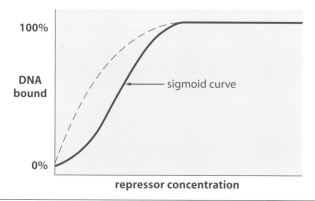

Figure 4.12. Binding of repressor to DNA. The dashed line is an example of an ordinary so-called Michaelis-Menten curve. It would describe the binding reaction if the repressor dimer were stable at all concentrations, and if a single dimer were to bind the operator site $O_R 1$. The sigmoid curve is that actually observed for the λ binding reaction. [Strictly speaking, the "repressor concentration" should be the concentration of free repressor (R), that is, the total repressor (R_T) minus that bound to operator (O_R). But in this experiment the repressor is in large excess over operator, and so (R) ~ (R_T).]

DNA concentration where all the repressors are bound, the number of DNA molecules that stick to the filter gives a measure of the number of active (DNA-binding) repressor molecules in a given preparation.

Figure 4.12 shows the binding of a DNA molecule bearing a single operator site—$O_R 1$—to repressor at various repressor concentrations, assayed using filters. For every point on this curve the mixture of repressor and DNA was allowed to reach equilibrium before filtering. The striking characteristic of the curve is that it is S shaped or sigmoid. This is in contrast to what we would expect if, for example, a single repressor monomer, bound to the operator, were sufficient to bind the DNA to the filter. In that case the shape of the curve would mimic that of the dotted line in the figure.

Let us consider for a moment the dotted curve of Figure 4.12. This curve would correspond to the following simple reaction:

$$RO \rightleftharpoons O + R \qquad K = \frac{(O)(R)}{(OR)} \tag{1}$$

where (R) = concentration of repressor, (O) = concentration of operator, and K is the equilibrium constant describing the reaction. (Note that here as throughout this book we write equilibrium constants as *dissociation constants*.) The value of K corresponds to the repressor concentration (R) at the midpoint of the curve, that is, where (O) = (O_R).

Why is the curve describing the interaction of λ repressor with its operator sigmoid? Rather than proceeding in a single step as in Equation (1), in this case the reaction proceeds in two steps. The first is dimerization of two repressor monomers, and the second is the binding of the dimer to the operator:

Dimerization:

$$R_2 \rightleftharpoons 2R \qquad K_1 = \frac{(R)^2}{(R_2)} \tag{2}$$

Operator binding:

$$R_2O \rightleftharpoons R_2 + O \qquad K_2 = \frac{(R_2)(O)}{(R_2O)} \tag{3}$$

The overall reaction can thus be described as:

$$R_2O \rightleftharpoons 2R + O \qquad K = K_1K_2 = \frac{(R)^2(O)}{(R_2O)} \tag{4}$$

where (R) = concentration of repressor monomer, and (R_2) = concentration of repressor dimer.

According to this picture, λ repressor dimers, which bind tightly to DNA, readily dissociate into monomers, which bind only very weakly. Thus at low concentration repressor is inactive in binding to DNA because it is predominantly monomer. As the concentration of repressor is increased, the monomers associate to form dimers and bind the operator. Equation (4) predicts that a plot of $ln(R_2O)/(O)$ versus lnR from the experiment of Figure 4.12 should yield a straight line of slope 2. It does, as shown in Figure 4.13.
References: 1, 4, 55.

Dimers form before DNA binding

We call the experiment shown in Figure 4.14 the "order of addition" experiment. Repressor is diluted, say 100-fold, from a concentrated solution into two tubes, the first of which contains labeled operator DNA, the second of which does not. After

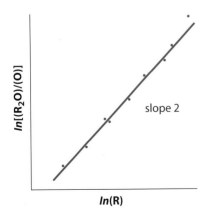

slope 2

Figure 4.13. Analysis of the sigmoid binding curve. Equation [4] rearranges to give $(R)^2/K = (R_2O)/(O)$. Thus a plot of $ln(R)$ vs. $ln[(R_2O)/(O)]$ should have a slope of 2, as it does. Were a tetramer the binding species, the slope would be 4. This graph is simply a replot of the sigmoid curve of Figure 4.12.

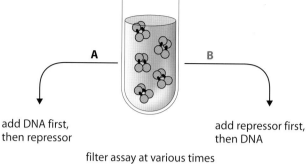

add DNA first,
then repressor

add repressor first,
then DNA

filter assay at various times

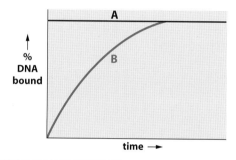

Figure 4.14. The "order of addition" experiment. Repressor monomers bind slowly to DNA because they must first form dimers, a slow process if the monomer concentration is low. Dimers bind quickly to DNA. If DNA is already present as in (A), dimers bind DNA before they fall apart.

a few moments labeled operator DNA is then added to the second tube, and the mixtures filtered to assay for repressor-operator complexes.

Strikingly different results are obtained in the two cases. The DNA added before addition of repressor is largely bound, whereas that added after repressor is largely unbound. If we continue to sample the two tubes we find that, slowly (in this case, over an hour or so), the DNA added after repressor also becomes bound.

We interpret these results as follows. At its original concentration, a significant fraction of repressor is dimer. When diluted into a solution containing DNA, the dimers bind operator quickly, before they can dissociate to monomers. But if DNA is absent the dimers dissociate to monomers until the concentration of dimers reaches the much lower equilibrium value characteristic of the more dilute solution. When DNA is now added, repressor binds slowly to the operators, the rate-limiting step being the formation of the dimeric intermediate.

Measurements of the monomer-dimer equilibrium of Equation (2) give a value of $K_1 = 2 \times 10^{-8}$M. This number, taken with the amount of repressor required to half-maximally bind DNA in an experiment such as that of Figure 4.12, allows us to calculate the value of $K_2 = 3 \times 10^{-9} M$ for binding to O_R1 at conditions designed to approximate those found in the cell, that is, at 0.2M KCl and 37°C. Under these conditions the mean lifetime of a repressor-operator complex is seconds. At lower temperatures or lower salt concentrations, at which the binding is tighter, the lifetime of the complex is minutes and even hours.

References: 1, 4, 51.

The amino domains contact DNA

A very useful way to study protein-DNA interaction is called "footprinting." Figure 4.15 illustrates the principle. A protein bound to a specific site on DNA protects that site from cleavage by (for example) DNase, an enzyme that otherwise randomly nicks the DNA. In practice one usually starts with a piece of DNA labeled with ^{32}P at the end of one of the strands of the duplex. DNase is added so that each molecule is nicked no more than once, and the DNA is then denatured and subjected to electrophoresis through a gel.

A "ladder" is formed in the gel, each successive rung of which (starting from

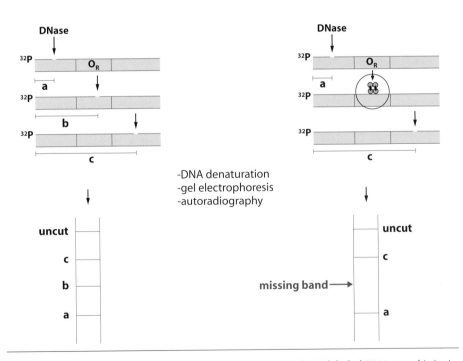

Figure 4.15. A DNase footprint experiment. In this experiment the unlabeled DNA strand is invisible, because the final products are visualized by autoradiography. In an actual experiment many bands appear, each arising from a labeled DNA molecule one base longer than the DNA in the band immediately below on the ladder.

the bottom) represents a strand of DNA that was nicked one base farther from the labeled end. If the experiment is now repeated in the presence of repressor, for example, one finds a gap (or "footprint") in the ladder corresponding to the positions on the DNA occupied by the repressor. That is, where repressor is bound DNase cannot nick and so fragments that would be generated by cutting the naked DNA at these sites are missing.

A related method for studying protein-DNA interaction uses the methylating agent dimethyl sulfate (DMS). DMS adds a methyl group primarily to two bases in double-stranded DNA, G and A. G is methylated at a position exposed in the major groove (nitrogen 7 in the standard numbering system) and A at a position exposed in the minor groove (nitrogen 3).

Each methylation sensitizes the DNA to cleavage. Breakage of molecules at the modified sites produces a ladder on a gel, each rung of which corresponds to the position of a methylated A or G residue. A bound repressor can protect bases from methylation and thus indicate base pairs that are closely approached by the bound protein. Occasionally a protein will, for unknown reasons, enhance the rate of methylation of a particular site, or indeed the rate of DNase cleavage of a particular site.

In footprinting experiments using DNase, repressor protects the 17 base pair operator plus a few bases on either side. As can be seen from Figure 4.16, repressor also protects G residues within the operator, but not those outside the operator,

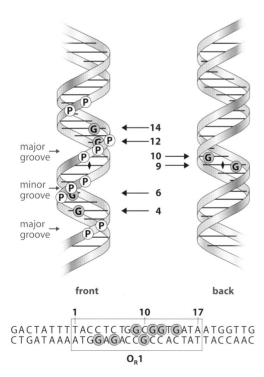

Figure 4.16. Positions on the operator contacted by repressor. In the sequence at the bottom the G residues protected by repressor from methylation by dimethyl sulfate are circled. These protected G's are displayed on the DNA helix in the figure. Four of the contacted G's are visible in the major groove on the frontside of the helix. The G at position 11 is hidden but, like the other G's on this side of the helix, it would be contacted by an λ-helix lying in the major groove along the front of the helix. In contrast, the site of repressor's contact with the G's at the two remaining positions is on the backside of the helix. The figure also shows the phosphates contacted by repressor. The diamond is the axis of twofold symmetry of the operator.

from methylation by DMS. The methylation of A residues is unaffected, consistent with the idea that repressor fills the major but not the minor groove of the operator.

In the present context, the important fact is that the separated amino domains behave identically to the intact dimer except that higher concentrations of the amino domain are required for the protection. Thus, the only contribution of the carboxyl domain to binding of a dimer to a single operator site is to strengthen the binding by holding together two amino domains. A second experiment indicates that in the cell, amino domains, in the absence of carboxyl domains, bind operator: a strain producing large amounts of a truncated repressor including only residues 1–92 is immune to super-infection by λ phages.

References: 23, 45, 58.

There are three 17 base pair repressor binding sites in the right operator. At each site repressor and Cro bind along the same face of the helix (Figures 1.16 and 1.22).

Inspection of the sequence of O_R reveals three similar 17 base pair sequences, each displaying imperfect twofold rotational symmetry. The left operator O_L also bears three similar sites (Table 2.1).

Chemical probes

Repressor or Cro can bind to any one of these sites, separated from the others, as assayed by the filter binding or footprint techniques. Both proteins protect from methylation G residues that lie along one face of the helix at each site. These are shown on the left side of Figure 4.16. Repressor, but not Cro, also protects one or two G's on the back of the helix near the middle of each site. As we document later, these backside contacts are made by the flexible arms of λ repressor.

The chemical ethyl nitrosourea can be used to identify those phosphates along the DNA backbone that are contacted by a bound protein. Also shown in Figure 4.16 are the phosphates contacted by repressor at O_R1. They lie along one face of the helix. The phosphates contacted by Cro are a subset of these, including the inner six but not the outer four, as we move in either direction from the center of the operator. For both repressor and Cro the contacted phosphates are symmetrically arrayed about the central base pair of the operator.

(The experiment that uses ethyl nitrosourea to identify phosphate contacts is called an ethylation interference experiment. The experiment reveals the effect on protein binding of an ethyl group which is first added to phosphates at each of several positions along the DNA backbone. Although we may often say that a phosphate is contacted by a protein, if we are relying on the results of an experiment with ethyl nitrosourea, we really mean that ethylation of that phosphate prevents binding of the protein.)

References: 29, 30, 31, 32, 61.

Operator mutations

At many positions in the 17 base pair sequence, a base pair substitution can be made that decreases the ability of the site to bind repressor. λvir, the multiply mutant phages that can grow in the presence of λ repressor, bear mutations in various operator sites. The first λvir isolated bears mutations in O_L1, O_R1, and O_R2. By starting with this λvir and selecting for mutants that grow in the presence of levels of repressor higher than those found in a lysogen, one recovers mutants bearing additional changes in these and other operator sites.

Figure 4.17 shows some of the operator mutations found in O_R. Almost every position in the operator is a potential site of mutation that will decrease the binding of repressor and/or Cro. Mutations at sites 2, 4, and 6 have the strongest effect on repressor binding, and these are the positions whose identities are most highly conserved as shown in Table 2.2. The bases at the remaining positions are used by repressor and Cro to distinguish among the various operator sites (see Figure 2.11). Mutations that lie between two of the operator sites—$P_{RM}Up$-1 and $P_{RM}116$—have no effect on repressor or Cro binding.
Reference: 2.

Binding to supercoiled and linear DNA

Repressor binds equally tightly to O_R1 carried on linear molecules or on negatively supercoiled molecules. This result strongly suggests that repressor does not greatly change DNA's structure, an inference confirmed by crystallographic analysis of a repressor-operator complex (see below). Were repressor to unwind DNA one turn, for example, it would bind much more tightly to the supercoil than to the linear molecule.
Reference: 37.

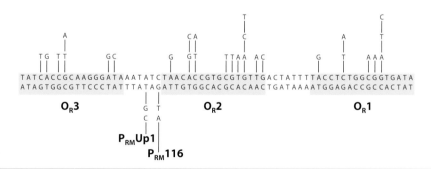

Figure 4.17. Mutations in and around O_R. This array of mutations should be compared with the sequences listed in Table 2.2. Mutations at position 3 decrease predominantly Cro but not repressor binding, and mutations at 8 and 9 decrease predominantly repressor but not Cro binding. Mutations at the remaining positions in the operator decrease the binding of both proteins. $P_{RM}Up$-1 renders P_{RM} active in the absence of repressor and $P_{RM}116$ inactivates P_{RM}.

Repressor binds to three sites in O_R with alternate pairwise cooperativity. The cooperativity is mediated by interactions between carboxyl domains of adjacent dimers (Figures 1.16, 1.17, 1.18, and 1.19).

The footprinting technique using DNase or DMS can be used to measure the affinity of repressor for each operator site. The experiment is performed at a variety of repressor concentrations in order to find that concentration at which one half of those sites are bound and thereby protected from chemical or nuclease attack. Cooperativity is revealed in this experiment by using DNA molecules bearing O_R with any one, two, or all three sites intact.

Figure 4.18 shows that the affinities of O_R1 and O_R2 are about equal on a wild-type template, even though their "intrinsic" affinities—those measured on DNA molecules bearing one or the other single site—differ more than tenfold. In other words, repressor binds cooperatively to O_R1 and O_R2 on a wild-type template. The affinity of site O_R3 is raised above its intrinsic affinity only if O_R1 is mutant. In that case repressors bind cooperatively to O_R2 and O_R3.

At O_L, repressor also binds according to the rule of alternate pairwise cooperativity to the three sites. As at O_R, O_L1 is the site with highest intrinsic affinity and O_L2 and O_L3 are about equal in intrinsic affinities. On a wild-type template, O_L1 and O_L2 bind equally tightly, and if O_L1 is mutant, O_L2 and O_L3 bind repressor cooperatively.

The separated amino domains bind to the three sites in O_R with an affinity order dictated by the intrinsic affinities of the sites. In other words, removal of the carboxyl domains has no effect on repressor's ability to distinguish the three sites, but it abolishes the cooperativity between dimers found at adjacent sites. The interaction energy between DNA-bound repressor dimers is about –2 kcal.

References: 10, 33.

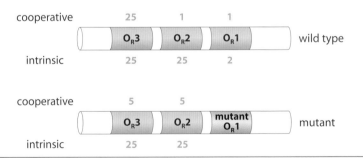

Figure 4.18. Relative affinities of operator sites for intact repressor. The numbers show the relative amount of repressor dimer required for half-maximal protection of each site in a footprint experiment. These numbers are therefore proportional to the dissociation constants. The affinities are different when the sites are artificially separated (the "intrinsic affinities") compared with when they are adjacent. The difference indicates cooperative binding to adjacent sites. No difference is observed if the separated amino domain of repressor is used instead of intact repressor.

In a lysogen repressor is typically bound to O_R1 and O_R2. The bound repressors turn off rightward transcription of *cro* and stimulate leftward transcription of *cl*. At higher concentrations, repressor binds to O_R3 to turn off transcription of *cl* (Figures 1.16 and 1.19).

This conclusion is based in part on a series of experiments that reveals the effects in vivo on P_R and on P_{RM} of a repressor bound to each operator site separately. These effects are shown in Figures 1.12 (repressor at O_R2), 1.13 (repressor at O_R1), and 1.14 (repressor at O_R3). Figure 4.19 shows the specially constructed bacterial strains that enabled us to perform this analysis.

We first isolated a segment of DNA bearing the λ promoters and then attached to either end of it a gene (*lacZ*) whose own promoter had been deleted. The product of *lacZ*, β-galactosidase, is easily assayed. In one construct *lacZ* transcription was driven by P_{RM} (Cell A in the figure), and in the other by P_R (Cell B). In the prototype constructs O_R was wild type, but in other cases only one or two functional repressor binding sites were present, the other(s) having been eliminated by mutation.

Both of the cells of Figure 4.19 also carried, on a plasmid, a source of λ repressor that provides a variable level of repressor. The plasmid bears the *cl* gene whose transcription is directed by the *lac* promoter (*Plac*). This DNA molecule also carries the *lac* repressor gene (*lacl*), whose product turns off the *lac* promoter. The amount of λ repressor made from this plasmid can be modulated by adding IPTG, a compound that inactivates the lac repressor, thereby relieving repression of the lac promoter and increasing synthesis of λ repressor.

In the ordinary situation repressor and Cro made under the control of P_{RM} and P_R, respectively, would feedback and alter promoter activity. In these experiments, however, β-galactosidase has no effect on either promoter, and so our strategy eliminates the complicating effects of feedback regulation. The experiment therefore reveals the effect of repressor on P_R and P_{RM} as it binds to the various sites in O_R, summarized as follows.

Figure 4.20 shows that low concentrations of λ repressor stimulate P_{RM} and high concentrations repress it. The figure also shows that λ repressor turns off transcription from P_R as it coordinately turns on transcription from P_{RM}, with half-maximal repression of P_R occurring at the repressor concentration at which P_{RM} is half-maximally activated. The repressor concentration at which P_{RM} is maximally activated is about that found in a lysogen. At this repressor concentration P_R is repressed over 99%.

What sites in O_R are occupied by repressor first to turn on P_{RM} and turn off P_R, and then to turn off P_{RM}? Experiments such as those of Figure 4.20 were repeated, but in these cases the λ operators bore mutations in two operator sites. These mutations abolish binding of repressor to the mutated sites, but leave the promoters themselves undamaged. Thus these experiments examine the effect of repressor bound, in each case, to only a single site.

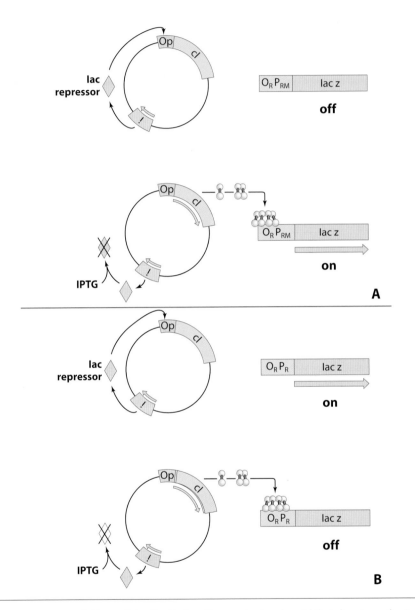

Figure 4.19. Two special bacterial cells. Both cells contain the plasmid that directs synthesis of λ repressor under control of lac repressor. Cell (A) also contains the P_{RM}-*lacZ* fusion and the other cell (B) the P_R-*lacZ* fusion. The more IPTG is added, the more λ repressor is made. The level of β-galactosidase in the cells grown at different IPTG concentrations shows how each promoter responds to different concentrations of λ repressor. Op stands for the *lac* operator, which, together with the *lac* promoter, has been placed in front of the *cI* gene.

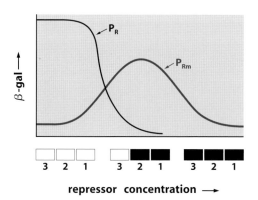

Figure 4.20. Regulation by repressor in vivo. This graph summarizes the results of an experiment using the two cells of Figure 4.19. For each case, β-galactosidase production was measured as a function of IPTG concentration. The midpoints of the repression and activation curves occur at the same IPTG concentration and hence at the same repressor concentration. The repressor level required to maximally activate P_{RM} is near that found in a lysogen. The boxes show the states of each operator site at the various repressor concentrations. When black they are occupied by repressor.

- $O_R2^-O_R3^-$. Repressor bound at O_R1 turns off P_R but does not activate P_{RM} (Figure 4.21A).

- $O_R1^-O_R2^-$. Repressor bound at O_R3 neither activates P_{RM} nor represses P_R (Figure 4.21B).

- $O_R1^-O_R3^-$. Repressor bound at O_R2 activates P_{RM} and represses P_R (Figure 4.21C).

The "Truth Table" (Table 4.2) summarizes the activities of P_R and P_{RM} when the operator contains a single repressor.

Analysis of the curves of Figure 4.21 reveals the effects of cooperative repressor binding to adjacent sites. For example, although repressor bound only to O_R2 activates P_{RM} (and represses P_R), more repressor is required to achieve the activation (and repression) in the case where O_R1 and O_R3 are mutant. The explanation is that in the mutant cases the helping effect of cooperativity is absent: repressor must bind to O_R2 unaided by interaction with repressor bound to O_R1.

A more dramatic effect of cooperativity is seen when the experiment of Figure 4.20 is performed using a template mutant in O_R1, but bearing wild-type sites O_R2 and O_R3. In this experiment the promoter fused to *lacZ* is a mutant, called $P_{RM}Up$-1, that works well in the absence of repressor. Using this mutant we can readily measure both the negative and the positive effects of repressor on P_{RM} function. Fig-

Table 4.2. The effect of a repressor dimer bound to a single site in O_R. In the absence of repressor, P_{RM} is off (unstimulated) and P_R is on, as shown in the top row.

P_{RM}	O_R3	O_R2	O_R1	P_R
off				on
off			X	off
on		X		off
off	X			on

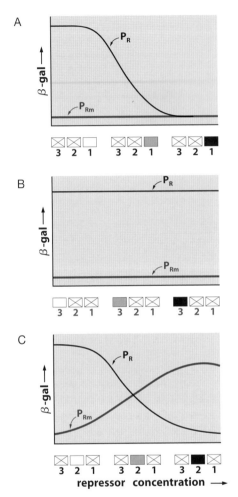

Figure 4.21. Regulation by repressor bound to a single site in vivo. These curves summarize the results of experiments similar to those described in Figure 4.20, except that here mutant operators are used. The curves describe the effect on P_R and P_{RM} of repressor bound to (A) O_R1, (B) O_R3, and (C) O_R2. The X indicates that the corresponding site is mutated so that repressor cannot bind there. When grey, some of the corresponding sites are occupied, and when black they are all occupied by repressor.

ure 4.22 shows the striking result: repressor, at a concentration only slightly higher than that which *activates* P_{RM} on a wild-type template, *represses* it on the O_R1^- template.

It is as though mutation of O_R1 has increased the repressor affinity of O_R3, but in fact the effect is due to alternate pairwise cooperativity. Thus, on the O_R1^- template repressor binds cooperatively to O_R2 and O_R3, and therefore O_R3 is filled at a concentration at which it would be free if the operator were wild type. Not surprisingly, this experiment also shows that repressor bound to O_R2 and O_R3 turns off P_R. Less repressor is required than if only O_R2 is intact, because of the helping effect of cooperative binding to O_R2 and O_R3.

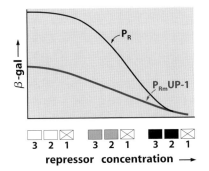

Figure 4.22. Alternate pairwise cooperativity in vivo. As for Figure 4.21, except that here we see the effect of repressor on P_R and $P_{RM}Up$-1 when bound at O_R2 and O_R3.

The results of these and additional experiments, performed in vivo, that display the effects of repressor bound to operators bearing one, two, or three functional sites, form a coherent picture when taken with the rules of repressor-operator interaction deduced from experiments performed in vitro (see Table 4.2). We conclude that in vivo there are two important states of O_R bearing repressor: in a lysogen repressor is bound primarily (cooperatively) to O_R1 and O_R2, thereby stimulating P_{RM} and repressing P_R. At higher concentration repressor binds to O_R3 (non-cooperatively) to turn off P_{RM} (Table 4.3).

(A note on mutations: some of the mutations used in the preceding and the following sections originally arose in operators of various λvir phages selected to grow lytically in the presence of repressor. Others were isolated from mutants identified using specially constructed phage and bacterial strains.)

Many of the effects of repressor observed in vivo can be reproduced in vitro. For example, purified RNA polymerase transcribes P_R, and this transcription is blocked by repressor (or by Cro). Neither protein has any effect on polymerase prebound to P_R, consistent with the idea that these proteins compete with polymerase for binding. At a concentration sufficient to fill O_R1 and O_R2, repressor stimulates transcription from P_{RM} and, at higher concentrations, represses it. We return later to a discussion of repressor's activity as a positive regulator.

References: 39, 41, 42, 43.

Table 4.3. The two physiologically important states of λ's O_R bearing repressor. The top row, with repressor at O_R1 and O_R2, P_{RM} on (stimulated) and P_R off describes the most common scenario found in lysogens.

P_{RM}	O_R3	O_R2	O_R1	P_R
on		X	X	off
off	X	X	X	off

Cro binds first to O_R3, then to O_R1 and O_R2, thereby first turning off P_{RM}, then P_R (Figure 1.23).

Some background about Cro

One of the earliest indications of Cro's existence was the result of the following genetic experiment (Figure 4.23). Starting with a λ-lysogen bearing a heat-sensitive mutation in its *cI* gene, most of the genes to the left of *cI* and to the right of *cro* were deleted. The presence of an intact *cI* gene was ensured by the fact that, at low temperature (30°C), the cells were immune to infection by λ phage.

When grown at 42°C, a temperature at which the mutant repressor was inactive, the cells displayed an unexpected phenotype: not only were they not immune (the expected result), they were "anti-immune." That is, wild-type λ, which could not grow on this strain at 30°C, formed only clear plaques on a lawn of this strain at 42°C. The effects were immunity specific; $λimm^{434}$, for example, formed normal turbid plaques on the strain at either temperature.

The inescapable conclusion was that the residual phage genome encoded a factor, specific for the λ-immunity region, that channeled infecting λ phages towards lytic growth. The factor was variously named Ai (anti-immunity); Tof (turn off) and Cro (control of repressor and other genes), based on this and related experiments. We now know that a single *cro* gene, transcribed from P_R, produces sufficient Cro to render a cell anti-immune. The anti-immune phenotype is evidently a consequence of partial repression by Cro of P_L and P_R and hence diminished expression of *cII* and *cIII*.

(Consider the state of the right operator-promoter in these anti-immune cells. As Cro binds it turns off P_R, but as the cells grow and divide, the concentration of Cro drops, and P_R turns on again. A steady state is reached at which the rate of synthesis of Cro just balances its rate of dilution and, presumably, a constant concen-

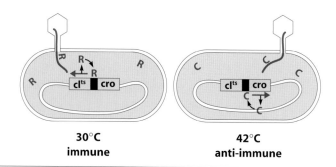

30°C
immune

42°C
anti-immune

Figure 4.23. Immunity and anti-immunity. The repressor (R) in these cells is a mutant that works well at 30°C, but denatures at 42°C. Cro (shown by C) is turned on at the high temperature, but the cell does not lyse because all the other phage genes have been deleted. Incoming phages are channeled by Cro to lytic growth. If the bacteria are shifted back to 30°C they occasionally switch back to the immune state.

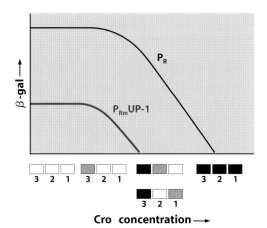

Figure 4.24. Regulation by Cro in vivo. Cro can be inserted in place of *cl* in the bacteria diagrammed in Figure 4.19. Then, addition of IPTG elicits synthesis of Cro. Using a P_R-*lacZ* fusion and a $P_{RM}Up$-1 *lacZ* fusion we see that Cro has a negative effect on both λ promoters.

tration of Cro is maintained. In this situation Cro diminishes [turns down], but does not abolish, its own synthesis.)

Cro was originally isolated from lytically infected cells as a protein that specifically bound λ DNA. The gene has been cloned and fused to *lac* promoter derivatives, and cells containing plasmids carrying these fusions can be directed to produce various levels of Cro. The protein is easily isolated from cells producing it in large amounts (several percent of their total protein as Cro) and many of the biochemical studies performed with repressor have also been performed with Cro.

References: 19, 20, 22, 31, 47, 44, 56.

Cro in vivo

Cells bearing plasmids producing Cro under control of *lac* repressor have been used to study the effect of various levels of Cro on P_{RM} and P_R in vivo. The results of an experiment analogous to that of Figure 4.20 shows that Cro turns off P_{RM} ($P_{RM}Up$-1 is used here) and at higher concentrations, it also turns off P_R (Figure 4.24). Studies of the effects of Cro using mutant operators, analogous to the repressor experiments, reveal the effect of Cro bound singly to each of the sites in O_R (Table 4.4).

Table 4.4. The effect of a Cro dimer bound to a single site in O_R. In the absence of Cro or repressor (top line) P_{RM} as well as P_R is on because the mutant $P_{RM}Up$-1 is used.

$P_{RM}Up$-1	O_R3	O_R2	O_R1	P_R
on				on
on			X	off
on		X		off
off	X			on

Table 4.5. The physiologically important states of λ's O_R bearing Cro.

$P_{RM}Up$-1	O_R3	O_R2	O_R1	P_R
on				on
off	X			on
off	X	X		off
off	X		X	off
off	X	X	X	off

Thus, as a negative regulator Cro acts identically to repressor; at O_R1 or O_R2 it turns off P_R, and at O_R3 it turns off P_{RM}. One important difference is that, unlike repressor, Cro bound at O_R2 does not stimulate P_{RM}. This information, taken with the order of binding to the sites in O_R1 (see Figure 1.23), shows that the physiologically important states of operator bearing Cro are as in Table 4.5. The table also shows the effect of Cro in these states.

Reference: 42.

Cro in vitro

Cro forms a stable dimer, and it shows no evidence of having more than one domain, a fact confirmed by crystallography as shown in Figure 2.9. Footprinting experiments show that Cro binds non-cooperatively to the three sites in O_R and to the three sites in O_L. Cro's affinity order for the O_R sites is as shown in Figure 1.23 ($O_R3 > O_R2 = O_R1$) and that for the O_L sites is $O_L1 > O_L2 = O_L3$. At 0.2M KCl at 37°C, Cro binds as tightly to O_R3 as a repressor dimer binds to O_R1. Cro's affinities for O_R2 and O_R1 are about tenfold lower than that for O_R3.

Cro blocks transcription of P_R. Sites O_R1 or O_R2 must be bound by Cro for this repression, but elimination of site O_R3 has no effect on repression of P_R. Cro also blocks transcription of P_{RM}, on a wild-type template bearing repressors at O_R1 and O_R2, and from the $P_{RM}Up$-1 mutant template. In both cases, this repression is abolished if O_R3 is mutant so that Cro cannot bind there.

References: 32, 62, 63.

RecA cleaves repressor to trigger induction (Figure 1.21).

Following induction of a wild-type lysogen, cleaved repressor can be detected in cell extracts. The cleavage reaction requires ultraviolet irradiation and the product of the *recA* gene. If the host bears an inactivating lesion in its *recA* gene, repressor is not cleaved and no phage induction is achieved. Certain mutant forms of *recA* can bypass the need for ultraviolet irradiation under some conditions: they will induce lysogens when the cells are grown at high temperatures but not irradiated.

The cleavage reaction can be carried out in vitro using purified RecA and repressor. Lambda repressor is cleaved between the Ala and Gly residues at positions 111 and 112, in a reaction that is stimulated by short single-stranded DNA fragments and ATP.

Several other repressors are also cleaved in vitro including the 434 repressor and LexA. In these cases also, the cleavage occurs at an Ala-Gly bond in the peptide that connects the two domains. These repressors cleave spontaneously at the Ala-Gly bond at pH 10.0, and more slowly at pH 7.0. Purified RecA greatly increases the rate of cleavage at the lower pH.

Cleavage of 30% of the repressor in a lysogen (assayed by filter binding) does not suffice to cause efficient induction, but cleavage of about 80% of repressor does.

References: 16, 36, 57, 59.

When Cro is bound at O_R3 the switch is thrown (Figure 1.24).

Two experiments show that Cro binding to O_R3 plays a critical role in induction are as follows.

The first experiment examines a λ phage bearing a mutant O_R3 to which Cro cannot bind. This phage forms turbid plaques of ordinary appearance, but lysogens bearing this phage induce very poorly in response to ultraviolet irradiation. This means that Cro binding to O_R3 must be necessary for ordinary induction.

For the second experiment, a plasmid producing Cro under control of *lac* repressor was introduced into a wild-type lysogen in one case, and into an O_R3-mutant lysogen in the other. When Cro synthesis was induced by addition of IPTG, the wild-type lysogen, but not the O_R3-mutant lysogen, was induced. Thus the binding of Cro to O_R3 triggers induction.

Reference: 5.

Repressor and Cro bind to the operator as shown in Figures 2.6, 2.8, 2.10, and 2.11.

The pictures are meant to imply not only that the complexes have the indicated configurations, but also that amino acids along the recognition helices (as well as λ's arm) determine the sequence specificity of binding. These closely related claims are supported by experiments involving crystallography and biochemistry.

Crystallography

The crystal structures of a number of specific DNA-binding proteins from bacteria have been solved to high resolution. These include the catabolite activator protein (CAP) of *E. coli*, as well as the repressor and Cro proteins of phage λ and its close relative phage 434. In each of these structures we find a helix-turn-helix (bihelical) motif very similar to that described in Chapter Two. In each case, model building shows that this motif is positioned on the surface of the protein so that the second helix of the pair (corresponding to λ repressor's helix 3) can fit into the major groove of DNA with the preceding helix lying across that groove. A low-resolution structure of λ repressor (the amino domain) complexed with an operator site confirms this general picture.

References: 13, 14, 17, 40, 49.

The "helix swap" experiment

The recognition α-helix of the 434 repressor—that seen lying in the major groove in the repressor-operator complex—is an important, perhaps the sole, determinant of its DNA-binding specificity. This conclusion follows from an experiment in which the 434 repressor was redesigned and its DNA-binding specificity changed. We call the experiment a "helix swap," although it might be more accurate to call it a "helix redesign." The experiment is outlined in Figure 4.25.

The idea was to replace the residues along the outside of 434's recognition helix with the corresponding residues from a different repressor—that of *Salmonella* phage P22—while leaving unchanged the residues along the inside surface. (See Chapter Two for a description of the inside and outside surfaces of the recognition helix.) The hope was that the resultant hybrid stretch of amino acids would form a λ-helix that would fold normally with the rest of the body of the 434 repressor, and would interact with the P22 but not with the 434 operator. The ordinary 434 and P22 repressors have no affinity for each other's operators as measured in vitro or in vivo. A P22 operator site is 22 base pairs, and the sequence of P22 O_R1 is shown along with a 434 operator sequence in Figure 4.25.

The upper part of Figure 4.26A shows the amino acids as they are disposed along the surfaces of the 434 and P22 recognition helices, and the lower part of the figure shows the presumed disposition along the redesigned recognition helix. The 434 repressor was redesigned by replacing a portion of its gene with synthetic DNA designed to encode the new recognition helix. The modified 434 repressor gene was attached to a strong promoter, and cells bearing this construct on a plasmid produce large amounts of the protein.

The striking result of the experiment was that the new protein bound to P22 operators, but not to 434 operators as assayed in vivo and in vitro. Thus cells containing the protein were immune to P22 but not to 434 phages, and the purified protein bound to the three sites in P22's O_R with an affinity order indistinguishable from that with which wild-type P22 repressor binds. The experiment shows that the sole determinants of specificity that distinguish these two repressors lie along the outside of their recognition helices.

References: 9, 38, 52, 60, 65, 66.

A repressor heterodimer that binds to a hybrid operator

We expect that the modified 434 repressor created in the helix swap experiment just described would form heterodimers with wild-type 434 repressor. This follows

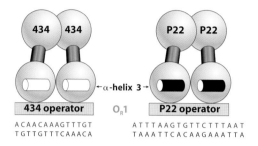

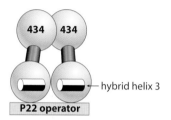

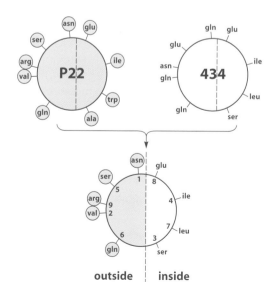

Figure 4.25. The "helix swap" experiment. In this experiment, amino acids along the outside of the recognition helix of 434 repressor are replaced with those from the corresponding position in the P22 repressor.

because all residues involved in dimerization by these two proteins are identical. As indicated in Figure 4.26B, a mixture of purified preparations of these two proteins contains a protein, presumably a dimer, that binds specifically to a hybrid operator that bears one 434 operator half-site fused to a P22 operator half-site. This result is as expected if each monomer of the heterodimer inserts its recognition helix into the major groove of one half-site of the operator.

Figure 4.26A. Recognition λ-helices involved in the helix swap experiment. In this representation we are looking down the barrels of three recognition helices. By comparing with Figure 2.3 we see that the R groups of amino acids along the "outside" surface point toward the DNA.

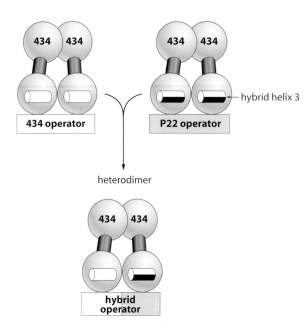

Figure 4.26B. Binding of a repressor heterodimer to a hybrid operator. Because the two half-sites of the hybrid operator are derived from different operators, the hybrid operator, lacks twofold rotational symmetry. [Hollis, M., Valenzuela, D., Pioli, D., Wharton, R., and Ptashne, M. (1988). Specific recognition of a chimeric operator by a repressor heterodimer. *Proc.* Natl. *Acad. Sci. USA* 85, 5834–5838.]

Specific amino acid-base pair contacts

All but one of the specific contacts shown in Figure 2.11 are evident in high-resolution crystallographic structures of repressor (amino domains)-DNA and Cro-DNA complexes, the references for which are given in Chapter Five. The exception is represented by the dashed arrow. In this case, the structure indicates that the Ala side chain cannot reach the indicated base pair, and the indirect contact is inferred from other kinds of experiments.

The contacts made by the conserved Gln-Ser residues (as well as some of the other contacts) were predicted from models built by docking the structure of each protein (solved without DNA) with models of the depicted operator sites. The following experiment provides further evidence that Ser at position 2 in the recognition helix recognizes position 4 in the operator. The sequences of the repressor and *cro* genes were changed so that they produced mutant proteins. Each of these mutants bears, at position 2 of the recognition helix, an alanine in place of the larger and chemically dissimilar serine. The mutant Cro and repressor proteins were purified for study in vitro.

The mutant and wild-type proteins were then compared in a methylation interference experiment. The experiment is similar in principle to those involving ethylnitrosourea discussed earlier. It is performed by first methylating operator DNA using DMS, adding one methyl group per DNA molecule, and then adding repressor or Cro. Operators that bind the protein are trapped on a nitrocellulose filter, and the methylated positions are identified by cleavage and gel electrophoresis (com-

pare with Figure 4.15). If methylation of a particular base blocks binding of the protein, the band in the gel corresponding to that operator position is absent.

The result was that methylation of the G at position 4 in the operator abolished binding of wild-type repressor and wild-type Cro, but had no effect on the
binding of the mutant repressor and the mutant Cro. Both mutant proteins bind
more weakly than do their wild-type counterparts, but methylation at position 4
has no effect on the binding of these mutants. In contrast, methylation of other
positions in the operator abolished binding of both the mutant and the wild-type
proteins.

References: 6, 18, 28, 46, 61.

The role of the arm of λ repressor

Experiments measuring protection from DMS-mediated methylation first indicated
that the arm of λ repressor makes specific contacts to G residues in the major
groove on the "backside" of the helix (Figure 4.16). At O_R1, repressor (or amino
domain) protects seven G residues from methylation. The N7 position of five of
these are exposed in the major groove on the "frontside" of the helix, but two,
those at positions 8 and 9, are exposed in the major groove on the backside.

Removal of residues 1–3 from intact repressor (by gene redesign) or from the
amino domain (by papain cleavage) produces repressors that protect the frontside
G's, but not the backside G's, from methylation. Removal of three terminal residues
decreases the binding to operator of otherwise intact repressor 30-fold, and
removal of six residues reduces the affinity by at least three orders of magnitude.

Genetic experiments also indicate that the arm of repressor makes at least one
specific contact on the back of the operator. Thus, a mutation at base 8 of O_R1
decreases binding affinity of repressor bearing an intact arm, but has no effect on
a repressor missing the amino terminal three residues. Thus the arm contributes
both to the energy and to the specificity of binding. NMR studies of free (that is,
not bound to operator) repressor indicate that the arm is flexible.

References: 21, 48, 64.

Repressor activates transcription of *cl* by binding to O_R2 and contacting poly-merase with its amino domain (Figures 1.12, 1.19, and 2.17).

Positive control mutants

Two general models might be proposed to explain how a DNA-bound regulatory
protein stimulates transcription. The first would hypothesize that the protein some-
how changes the conformation of DNA to which it binds, and that this changed
conformation would provide a more suitable substrate for RNA polymerase. The
second and contrasting model would imagine that the bound protein stimulates
transcription by contacting polymerase. The fact that crystallographic studies confirm our surmise that λ repressor changes DNA structure very little upon binding is

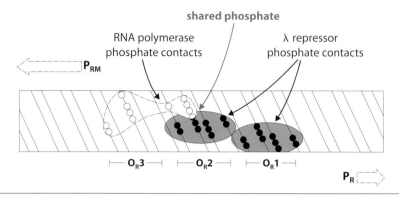

Figure 4.27. Arrangement of λ repressor at O_R1 and O_R2 and polymerase at P_{RM}. The DNA helix has been flattened to display all of the phosphates that are contacted by polymerase at P_{RM} and repressor at O_R1 and O_R2. (DNase footprint experiments show that polymerase covers a larger segment of DNA than is indicated by these phosphate contacts. Throughout the book we have pictured polymerase as covering the larger region.)

an argument against the first model. A compelling argument against that model is provided by the properties of repressor mutants called *pc*, for *positive control*.

These *pc* repressor mutants bind DNA normally but fail to activate transcription; they have been isolated for both λ and P22 repressors. In each case the mutations change residues that, on the basis of structural and other studies, are predicted to closely approach (touch) polymerase. The mutants provide additional evidence, therefore, that our picture of the repressor-operator complexes in these two cases is correct.

Figure 4.27 shows the phosphates inferred to be contacted by polymerase bound at P_{RM} and by λ repressor bound at O_R2 and O_R1. One phosphate—the "shared" phosphate—is contacted by both λ repressor (at O_R2) and by polymerase when these proteins are analyzed separately, and is therefore likely to be near the interaction site between the two proteins. If we position a repressor dimer at O_R2 we find that the changed residue on each pc mutant λ repressor is on a patch of the surface that, according to this "shared" phosphate argument, is in closest proximity to polymerase (see Figure 2.17).

Figure 4.28 shows that the λ*pc* mutations lie in the bihelical DNA recognition unit, affecting residues in one helix and in the bend, but not residues in the recognition helix. This fact, taken with the following considerations, enables us to make a strong prediction. Figure 4.29 shows the position of P22 repressor at its O_R1 and O_R2, and polymerase at P_{RM}, as indicated by contacted phosphates. Note that whereas λ repressor and polymerase form a "train" along the DNA, or nearly so, P22 repressor and polymerase "sandwich" the DNA. Note also that a single phosphate contacted is "shared" by P22 repressor and polymerase.

Considering only the pattern of phosphates contacted by polymerase, the

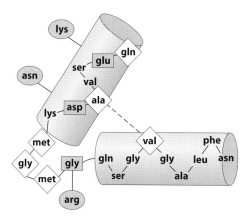

Figure 4.28. Lambda *pc* mutations. The amino acids that replace the wild-type residues in three λ *pc* mutants are circled. The residues in diamonds are those conserved in many bihelical units of specific DNA-binding proteins. Note that the *pc* mutations affect neither the conserved residues nor residues in the recognition helix.

shared phosphate is the same in the P22 case as at λ's P_{RM}. But considering repressor, the shared phosphate is in a different position from that in the λ case. We do not know the structure of P22 repressor, but we do know where its presumed bihelical DNA recognition unit lies along its sequence. It is evident from Figure 4.30 that if P22 repressor touches polymerase with any part of this bihelical unit in the vicinity of the shared phosphate, it must be with residues at or near the carboxyl end of the recognition helix. This is in clear contrast to the λ case, also shown on the figure, where the pc residues lie near the amino terminus of the recognition helix.

Some *pc* mutations of P22 repressor lie at the very carboxyl end of the presumed recognition helix, and others probably (as deduced from arguments of structural homologies) lie nearby. The results suggest that these positive regulatory pro-

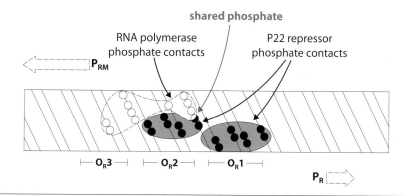

Figure 4.29. Arrangement of P22 repressor at O_R1 and O_R2 and polymerase at P_{RM}. Compare this figure, describing the contacted phosphates at P22's O_R and P_{RM} *with the* λ contacts in Figure 4.27.

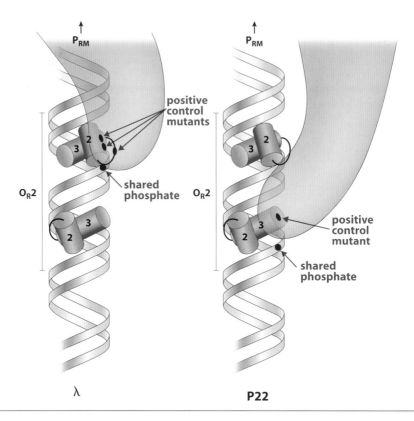

Figure 4.30. Relationship of "shared phosphate" to the bihelical units of repressor in λ and P22. As predicted from their different positions in relation to polymerase at P_{RM}, λ repressor and P22 repressor *pc* mutations arise on different surfaces of the two repressors.

teins might interact with the same surface of polymerase by using "different" repressor surfaces (see Figure 4.30). It may be significant that of the five *pc* mutants of λ and P22 so far isolated, each causes a change to an amino acid that is more basic (more positively charged) than that found at the corresponding position in the wild-type protein.

Studies with the purified proteins show that they bind DNA normally, or nearly so, and repress P_R, but fail to activate P_{RM}.

References: 24, 25, 27, 52.

Positive control in vitro

Repressor stimulates transcription from P_{RM} by polymerase some tenfold. This stimulation is readily observed only if the polymerase concentration is below a certain level; at high polymerase concentrations P_{RM} functions at a high level with or without repressor. This is as expected if repressor's role is to increase the rate at which polymerase interacts with P_{RM}.

Maximal stimulation of P_{RM} by repressor is observed at repressor concentrations that, as revealed by footprint experiments, are just sufficient to occupy O_R1 and O_R2. At higher repressor concentration O_R3 is also filled and P_{RM} is repressed. No repressor stimulation of P_{RM} is observed if the promoter is mutant in O_R1 and O_R2, or if it is mutant in O_R2 only. If O_R1 and O_R3 are mutant, P_{RM} can be stimulated, but higher concentrations of repressor are required than if the template bears an intact O_R1 as well. (Recall that O_R2 binds repressor more weakly if O_R1 is mutant.) The separated amino domain of repressor also stimulates P_{RM} in vitro and in vivo; in this case the effect is the same whether O_R1 is mutant or wild type. (Recall that the amino domain binds non-cooperatively.)

The results of the experiments of the preceding paragraphs are consistent with the idea that repressor at O_R2 stimulates P_{RM} by contacting polymerase bound there. But the following objection arises: perhaps polymerase at P_R interferes with polymerase binding to P_{RM}, and repressor stimulates P_{RM} merely by preventing binding of polymerase to P_R. This "competing polymerase" model is excluded by several kinds of experiments. For example, removal of P_R by deletion does not suffice to activate P_{RM}, and indeed has no effect on the response of P_{RM} to polymerase as assayed in vitro or in vivo. To take another example, P_{RM} cannot be significantly stimulated if the template is O_R2^-.

Repressor bound to O_R1 prevents binding of polymerase to P_R on this template, but this has no effect on P_{RM}. Moreover, on a template deleted for P_R and O_R1, polymerase bound at P_{RM} increases the affinity of O_R2 for intact repressor and for the separated amino domain.

Reference: 43.

CONCLUSION

The preface to this book noted that many of our pictures are not simple representations of single observations. Rather, particularly for the pictures of Chapter One, each summarizes and interprets a variety of experiments. The reader who has digested Chapter Four will readily appreciate this point.

She will realize, for example, that we have had to construct pictures of intact repressor monomers and dimers. Does the monomer really look like a dumbbell? Are we correct in picturing repressor dimers "leaning" to touch other dimers and to effect cooperative binding? Will flexibility of the linker between domains of repressor explain the rather surprising fact that, despite differences in spacing of several base pairs between adjacent sites in O_R and O_L, the energy of interaction between repressor dimers is about the same in all cases ($\sim$–2 kcal)? We suspect that interactions between DNA-bound regulatory proteins play an important role in gene control in all organisms, and we need to know more about the mechanisms involved.

If our knowledge of repressor structure is incomplete, our knowledge of RNA polymerase structure is virtually nil. To say that repressor acts as an activator of transcription by protein-protein contact is merely to begin to understand the mechanism. Do all activators in *E. coli* contact the same part of polymerase? How do these contacts help polymerase bind and begin transcription? These questions take on wider interest in view of the recent findings that the amino acid sequences of two of the subunits of *E. coli* RNA polymerase are significantly homologous to those of subunits of eukaryotic RNA polymerases (see Reference 11).

As noted in Chapter Three, we are only beginning to understand in detail how regulatory proteins recognize specific base sequences in DNA. Although we can describe the functions of λ's N and Q proteins, we know virtually nothing about their structures and therefore very little about the detailed mechanisms involved. We know the proteins involved in other aspects of λ growth integration—and excision, for example, and DNA replication and recombination—but here again our knowledge of mechanisms is just emerging. If our goal is to understand these processes in terms of molecular structure, we must regard their study as having only begun.

FURTHER READING: REVIEW ARTICLES

1. Chadwick, P., Pirrotta, V., Steinberg, R., Hopkins, N., and Ptashne, M. (1970). The λ and 434 phage repressors. *Cold Spring Harbor Symp. Quant. Biol.* 35, 283–294.

2. Gussin, G., Johnson, A., Pabo, C., and Sauer, R. (1983). Repressor and Cro protein: Structure, function, and role in lysogenization. In *Lambda II*, R.W. Hendrix, J.W. Roberts, F.W. Stahl, and R. Weisberg, eds. (New York: Cold Spring Harbor), pp. 93–123.

3. Jacob, F. and Monod, J. (1961). Genetic regulatory mechanisms in the synthesis of proteins. *J. Mol. Biol.* 3, 318–356.

4. Johnson, A.D., Pabo, C.O., and Sauer, R.T. (1980). Bacteriophage λ repressor and Cro protein: Interaction with operator DNA. *Methods in Enzymology* 65, 839–856.

5. Johnson, A.D., Poteete, A.R., Lauer, G., Sauer, R.T., Ackers, G.K., and Ptashne, M. (1981). λ repressor and Cro-components of an efficient molecular switch. *Nature* 294, 217–233.

6. Lewis, M., Jeffrey, A., Wang, J., Ladner, R., Ptashne, M., and Pabo, C.O. (1983). Structure of the operator-binding domain of λ repressor: Implication for DNA recognition and gene regulation. *Cold Spring Harbor Symp. Quant. Biol.* 47, 435–440.

7. Lwoff, A. (1953). Lysogeny. *Bact. Rev.* 17, 269–337.

8. Miller, J.H. (1978). The *lasI* gene: Its role in *lac* operon control and its use as a genetic system. In *The Operon*, J.H. Miller and W.S. Reznikoff, eds. (New York: Cold Spring Harbor), pp. 31–89.

9. Wharton, R.P. and Ptashne, M. (1986). An λ-helix determines the DNA-binding specificity of a repressor. *Trends in Biochemical Sciences* 11, 71–73.

RESEARCH ARTICLES

10. Ackers, G.K., Shea, M.A., and Johnson, A.D. (1982). Quantitative model for gene regulation by λ phage repressor. *Proc. Natl. Acad. Sci. USA* 79, 1129–1133.

11. Allison, L.A., Moyle, M., Shales, M., and Ingles, C.J. (1985). Extensive homology among the largest subunits of eukaryotic and prokaryotic RNA polymerases. *Cell* 42, 599–610.

12. Amann, E., Brosius, J., and Ptashne, M. (1983). Vectors bearing a hybrid tip-lac promoter useful for regulated expression cloned genes in *E. coli. Gene* 25, 167–178.

13. Anderson, W.F., Ohlendorf, D.H., Takeda, Y., and Matthews, B.W. (1981). Structure of the *cro* repressor from bacteriophage λ and its interaction with DNA. *Nature* 290, 754–758.

14. Anderson, J.E., Ptashne, M., and Harrison, S.C. (1985). The structure of a phage repressor-operator complex at 7 A resolution. *Nature* 316, 596–601.

15. Backman, K. and Ptashne, M. (1978). Maximizing gene expression on a plasmid using recombination *in vitro. Cell* 13, 65–71.

16. Bailone, A., Levine, A., and Devoret, R. (1979). Inactivation of prophage λ repressor *in vivo. J. Mol. Biol.* 131, 553–572.

17. Bushman, F.D., Anderson, J.E., Harrison, S.C., and Ptashne, M. (1985). Ethylation interference and X-ray crystallography identify similar interactions between 434 repressor and operator. *Nature* 316, 651–653.

18. Ebright, R.H. (1986). Proposed amino acid-base pair contacts for 13 sequence specific DNA-binding proteins. In *Protein Structure, Folding, and Design*, D. Oxender, ed. (New York: Alan R. Liss).

19. Echols, H., Green, L., Oppenheim, B., Oppenheim, A., and Honigman, A. (1973). Role of the *cro* gene in bacteriophage λ development. *J. Mol. Biol.* 80, 203–216.

20. Eisen, H., Brachet, P., Pereira da Silva, L., and Jacob, F. (1970). Regulation of repressor expression in λ. *Proc. Natl. Acad. Sci. USA* 66, 855–862.

21. Eliason, J., Weiss, M.A., and Ptashne, M. (1985). NH_2-terminal arm of phage λ repressor contributes energy and specificity to repressor binding and determines the effects of operator mutations. *Proc. Natl. Acad. Sci. USA* 82, 2339–2343.

22. Folkmanis, A., Takeda, Y., Simuth, J., Gussin, G., and Echols, H. (1976). Purification and properties of a DNA-binding protein with characteristics expected for the Cro protein of bacteriophage λ, a repressor essential for lytic growth. *Proc. Natl. Acad. Sci. USA* 73, 2249–2253.

23. Galas, D.J. and Schmitz, A. (1978). DNase footprinting: A simple method for the detection of protein-DNA binding specificity. *Nucl. Acids Res.* 5, 3157–3170.

24. Guarente, L., Nye, J.S., Hochschild, A., and Ptashne, M. (1982). A mutant λ repressor with a specific defect in its positive control function. *Proc. Natl. Acad. Sci. USA* 79, 2236–2239.

25. Hawley, D.K. and McClure, W.R. (1983). The effect of a λ repressor mutation on the activation of transcription initiation from the λP_{RM} promoter. *Cell* 32, 327–333.

26. Hecht, M.H., Nelson, H.C.M., and Sauer, R.T. (1983). Mutations in λ repressor's amino-terminal domain: Implications for protein stability and DNA-binding. *Proc. Natl. Acad. Sci. USA* 80, 2676–2680.

27. Hochschild, A., Irwin, N., and Ptashne, M. (1983). Repressor structure and the mechanism of positive control. *Cell* 32, 319–325.

28. Hochschild, A. and Ptashne, M. (1986). Homologous interactions of λ repressor and λ Cro with the λ operator. *Cell* 44, 925–933.

29. Humayun, Z., Jeffrey, A., and Ptashne, M. (1977). Completed DNA sequences and organizations of repressor-binding sites in the operators of phage λ. *J. Mol. Biol.* 112, 265–277.

30. Humayun, Z., Kleid, D., and Ptashne, M. (1977). Sites of contact between λ operators and λ repressor. *Nucl. Acids Res.* 4, 1595–1607.

31. Johnson, A. (1980). Mechanism of action of the λ Cro protein. Ph.D. thesis, Harvard University, Cambridge, Massachusetts.

32. Johnson, A., Meyer, B.J., and Ptashne, M. (1978). Mechanism of action of the Cro protein of bacteriophage λ. *Proc. Natl. Acad. Sci. USA* 75, 1783–1787.

33. Johnson, A.D., Meyer, B.J., and Ptashne, M. (1979). Interactions between DNA-bound repressors govern regulation by the λ phage repressor. *Proc. Natl. Acad. Sci. USA* 76, 5061–5665.

34. Kaiser, A.D. (1957). Mutations in a temperate bacteriophage affecting its ability to lysogenize *Escherichia coli. Virology* 3, 42–61.

35. Kaiser, A.D. and Jacob, F. (1957). Recombination between related temperate bacteriophages and the genetic control of immunity and prophage localization. *Virology* 4, 509–521.

36. Little, J.W. (1984). Autodigestion of LexA and phage λ repressors. *Proc. Natl. Acad. Sci.* USA 81, 1375–1379.

37. Maniatis, T. and Ptashne, M. (1973). Multiple repressor binding in the operators of bacteriophage λ. *Proc. Natl. Acad. Sci. USA* 70, 1531–1535.

38. Matthews, B.W., Ohlendorf, D.H., Anderson, W.F., and Takeda, Y. (1982). Structure of DNA-binding region of *lac* repressor inferred from its homology with *cro* repressor. *Proc. Natl. Acad. Sci. USA* 79, 1428–1432.

39. Maurer, R., Meyer, B.J., and Ptashne, M. (1980). I. O_R3 and autogenous negative control by repressors. *J. Mol. Biol.* 139, 147–161.

40. McKay, D., Weber, I., and Steitz, T. (1982). Structure of catabolite gene activator protein at 2.9-8f resolution. *J. Biol. Chem.* 257, 9518–9524.

41. Meyer, B.J., Kleid, D.G., and Ptashne, M. (1975). λ repressor turns off transcription of its own gene. *Proc. Natl. Acad. Sci. USA* 72, 4785–4789.

42. Meyer, B.J., Maurer, R., and Ptashne, M. (1980). II. O_R1, O_R2, and O_R3: Their roles in mediating the effects of repressor and cro. *J. Mol. Biol.* 139, 163–194.

43. Meyer, B.J. and Ptashne, M. (1980). III. λ repressor directly activates gene transcription. *J. Mol. Biol.* 139, 195–205.

44. Neubauer, Z. and Calef, E. (1970). Immunity phase-shift in defective lysogens: Non-mutational hereditary change of early regulation of λ prophage. *J. Mol. Biol.* 51, 1–13.

45. Ogata, R. and Gilbert, W. (1979). DNA-binding site of *lac* repressor probed by dimethylsulfate methylation of *lac* operator. *J. Mol Biol.* 132, 709–728.

46. Ohlendorf, D.H., Anderson, W.F., Fisher, R.G., Takeda, Y., and Matthews, B.W. (1982). The molecular basis of DNA-protein recognition inferred from the structure of *cro* repressor. *Nature* 298, 718–723.

47. Openheim, A.B., Neubauer, Z., and Calef, E. (1970). The antirepressor: A new element in the regulation of protein synthesis. *Nature* 226, 31–32.

48. Pabo, C.O., Krovatin, W., Jeffrey, A., and Sauer, R.T. (1982). The N-terminal arms of λ repressor wrap around the operator DNA. *Nature* 298, 441–443.

49. Pabo, C.O. and Lewis, M. (1982). The operator-binding domain of λ repressor: Structure and DNA recognition. *Nature* 298, 443–447.

50. Pabo, C.O., Sauer, R.T., Sturtevant, J.M., and Ptashne, M. (1979). The λ repressor contains two domains. *Proc. Natl. Acad. Sci. USA* 76, 1608–1612.

51. Pirrotta, V., Chadwick, P., and Ptashne, M. (1970). Active form of two coliphage repressors. *Nature* 227, 41–44.

52. Poteete, A.R. and Ptashne, M. (1982). Control of transcription by the bacteriophage P22 repressor. *J. Mol. Biol.* 157, 21–48.

53. Ptashne, M. (1967). Isolation of the λ phage repressor. *Proc. Natl. Acad. Sci. USA* 57, 306–313.

54. Ptashne, M. (1967). Specific binding of the λ phage repressor to λ DNA. *Nature* 214, 232–234.

55. Riggs, A.D., Suzuki, H., and Bourgeois, S. (1970). *lac* repressor-operator interaction. *J. Mol. Biol.* 48, 67–83.

56. Roberts, T.M., Kacich, R., and Ptashne, M. (1979). A general method for maximizing the expression of a cloned gene. *Proc. Natl. Acad. Sci. USA* 76, 760–764.

57. Roberts, J.W., Roberts, C.W., and Mount, D.W. (1977). Inactivation and proteolytic cleavage of phage λ repressor in vitro in an ATP-dependent reaction. *Proc. Natl. Acad. Sci. USA* 74, 2283–2287.

58. Sauer, R.T., Pabo, C.O., Meyer, B.J., Ptashne, M., and Backman, K.D. (1979). Regulatory functions of λ repressor reside in the amino-terminal domain. *Nature* 279, 396–400.

59. Sauer, R.T., Ross, M.J., and Ptashne, M. (1982). Cleavage of the λ and P22 repressors by RecA protein. *J. Biol. Chem.* 257, 4458–4462.

60. Sauer, R.T., Yocum, R.R., Doolittle, R.F., Lewis, M., and Pabo, C.O. (1982). Homology among DNA-binding proteins suggests use of a conserved super-secondary structure. *Nature* 298, 447–451.

61. Siebenlist, A. and Gilbert, W. (1980). Contact between *Escherichia coli* RNA polymerase and an early promoter of phage T7. *Proc. Natl. Acad. Sci.* 77, 122–126.

62. Takeda, Y. (1979). Specific repression of *in vitro* transcription by the Cro repressor of bacteriophage λ. *Mol. Biol.* 127, 177–189.

63. Takeda, Y., Folkmanis, A., and Echols, H. (1977). Cro regulatory protein specified by bacteriophage λ. *J. Biol. Chem.* 252, 6177–6183.

64. Weiss, M.A., Eliason, J.L., and States, D.J. (1984). Dynamic filtering by two-dimensional ^{1}H NMR with application to phage λ repressor. *Proc. Natl. Acad. Sci. USA* 81, 6019–6023.

65. Wharton, R.P., Brown, E.L., and Ptashne, M. (1984). Substituting an λ-helix switches the sequence specific DNA interactions of a repressor. *Cell 38,* 361–369.

66. Wharton, R.P. and Ptashne, M. (1985). Changing the binding specificity of a repressor by redesigning an λ-helix. *Nature* 316, 601–605.

CHAPTER FIVE

2004: NEW DEVELOPMENTS

E ach of the following five sections describes an aspect of the switch that, in whole or in part, has come to light since 1986. The first three of these lend themselves to an approach used in the original text: first a story is told (as in Chapters One through Three), and then the evidence is presented (as in Chapter Four).

---

1. LONG-RANGE COOPERATIVITY AND REPRESSION OF P_{RM}

Our original description of the switch emphasized interactions of two regulatory proteins—repressor and Cro—with a single region of DNA, the right operator (O_R). By binding to specific sites in this operator the proteins control the activities of two promoters.

Figure 19 of Chapter One (i.e., Fig. 1.19) shows an example: repressor bound at two sites in O_R activates transcription of its own gene (from the promoter P_{RM}) as it represses transcription of the *cro* gene (from the promoter P_R). Cooperativity plays a key role: a repressor dimer binding to the strong site O_R1 helps another dimer bind to the adjacent weak site O_R2 (see Fig. 1.16). The figure implies a second auto-regulatory effect as well: repressor can turn off transcription of its own gene by binding to the weak operator site O_R3.

As explained in Chapter Four, these activities of repressor were deduced mainly by studying the O_R region in isolation from the rest of the λ chromosome. This approach allowed us to identify the individual components of the switch and many of their interactions. We had no strong reasons to think our description was incomplete, but it was.

We now know that, in the context of the intact λ chromosome, repressor molecules bind cooperatively not just to adjacent sites, but to sites separated by over 2000 base pairs. And, we now know, the binding reaction thought to lack cooperativity—that of a repressor dimer to O_R3—is influenced in an important way by "long-range" repressor-repressor interactions. Thanks to these new interactions, the switch works even more efficiently than we had thought.

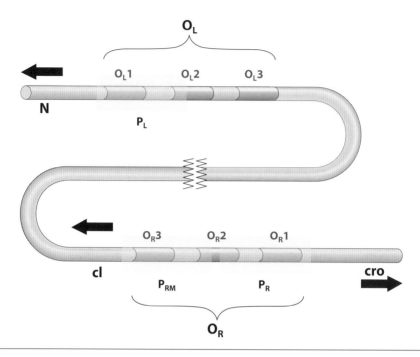

Figure 5.1. Disposition of O_R and O_L on the λ chromosome. The arrows show the direction of transcription emanating from each promoter. Additional genes transcribed from P_R (starting with *cro*) and from P_L (starting with *N*) are shown in Fig. 3.6.

In the following section we show how repressor binding to O_R is influenced by the presence of the second operator O_L (the "left" operator) positioned 2.4-kb away on the λ chromosome. As shown in Figure 5.1, O_L, like O_R, has three repressor binding sites. O_L1 (like O_R1 in O_R) is the site with the highest affinity for repressor. Just as at O_R, repressor binds cooperatively to two sites (in this case O_L1 and O_L2) and turns off transcription of an adjacent promoter (in this case P_L). There is no promoter associated with O_L analogous to P_{RM}, however, and so for the moment the role of O_L3 remains mysterious.

An Octamer of Repressor Binds O_R and O_L

On a DNA molecule bearing both O_L and O_R, repressor binds cooperatively to four sites—O_L1, O_L2, O_R1, and O_R2. The reaction requires that repressor form an octamer (four dimers) and that the DNA between O_L and O_R loop out as shown in Figure 5.2.

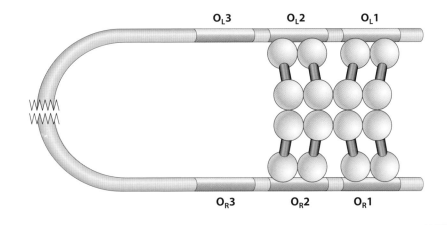

Figure 5.2. "Long-range" cooperativity. The repressor octamer contacts two sites in O_R and two in O_L, with the DNA in between forming a loop. P_R and P_L are turned off, and P_{RM} is turned on.

In an earlier discussion we saw how cooperative binding of repressor to two sites was predicted to create an "on-off" switch (see Fig. 1.25). Cooperative binding to four sites is expected to modestly increase this effect. As shown in Figure 5.3, the curve describing repression of P_R (or P_L) as a function of the repressor concentration is calculated to be somewhat steeper (more "sigmoid") than we had thought. As the repressor concentration drops, the activity of the promoter, at first buffered against the change, switches from "off" to "on" more precipitously than in the absence of the additional layer of cooperativity. And P_R and P_L, in a lysogen, are repressed to an even greater degree than previously appreciated. The following section describes a more dramatic effect of long-range cooperativity.

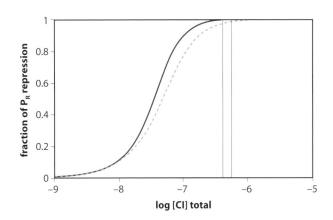

Figure 5.3. The effect of "long-range" cooperativity on repression of P_R. The curves depict repression of P_R, as a function of the repressor concentration, in the presence (*solid*) and absence (*dotted*) of O_L (i.e., in the presence and absence of the long-range interactions). The bar shows the concentration of repressor in a wild-type lysogen.

Autonegative Regulation of Repressor Synthesis

The original description of the switch underestimated the efficiency of repressor auto-negative regulation. As shown in Figure 1.15, we had deduced that, at the concentration typically found in lysogens, repressor bound O_R3 less than 10% of the time. We had surmised that repressor binding to O_R3 (and thus repression of P_{RM}) would become significant only were repressor levels to transiently increase at least tenfold. We had thought that repressor bound non-cooperatively to O_R3 (see Fig. 1.16), but we now know that—on a DNA molecule bearing both O_L and O_R—this is not true.

As shown in Figure 5.2, DNA looping (a consequence of repressor octamer binding to sites in O_R and O_L) brings O_R3 close to O_L3. As shown in Figure 5.4, this juxtaposition allows a repressor dimer binding to O_L3 to help another dimer bind to O_R3. This cooperative binding ensures that O_R3 is filled at a lower repressor concentration than it otherwise would be.

For example, in a lysogen, repressor binds O_R3 some 60–70% of the time, rather than the <10% previously estimated. As a result of this binding, the repressor concentration is maintained at a level some threefold lower than what it would be were repressor to never bind O_R3. Moreover, the system is poised so that any change—an increase or a decrease—in the repressor concentration in a lysogen results in a compensatory change in the rate of repressor synthesis. These effects are shown in Figure 5.5.

This newly recognized efficiency of autonegative regulation is crucial: without this control a lysogen induces poorly (in response to UV irradiation, for example) because it contains—by a factor of three—too much repressor. The role of O_L3 is no longer mysterious.

In the next section, I describe the key experimental findings that revealed the long-range interactions between repressors bound at O_R and O_L and their biological significance.

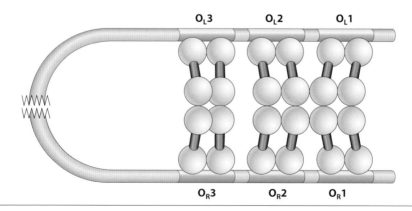

Figure 5.4. Binding of repressor tetramer to O_R3 and O_L3. In the presence of octamer bound at O_L1, O_L2, O_R1, and O_R2, two additional repressor dimers bind cooperatively, one to O_L3 and the other to O_R3. P_{RM} (as well as P_R and P_L) is turned off.

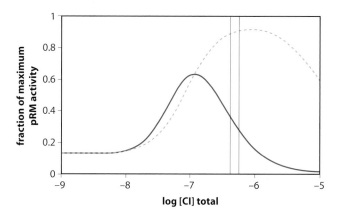

Figure 5.5. The effect of long-range interactions on activation and repression of P_{RM}. The curves depict the activity of P_{RM} in the presence (*solid*) or absence (*dotted*) of O_L. The bar shows the concentration of repressor in a wild-type lysogen. These curves, as well as those of Fig. 5.3, were calculated by K. Shearwin as described by Ackers et al. (1982) using additional data from Dodd et al. (2001 and 2004).

HOW DO WE KNOW

Long-range Interactions and Repression of P_R

The following two findings indicate that repressors binding to well-separated sites on the DNA interact with each other.

- Looped DNA structures form when repressor is added to DNA molecules bearing two widely separated pairs of operator sites (Fig. 5.6). The result suggests that repressors bound to the two pairs of sites interact with each other, with the intervening DNA looping out to accommodate the reaction.

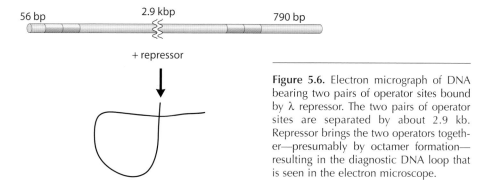

Figure 5.6. Electron micrograph of DNA bearing two pairs of operator sites bound by λ repressor. The two pairs of operator sites are separated by about 2.9 kb. Repressor brings the two operators together—presumably by octamer formation—resulting in the diagnostic DNA loop that is seen in the electron microscope.

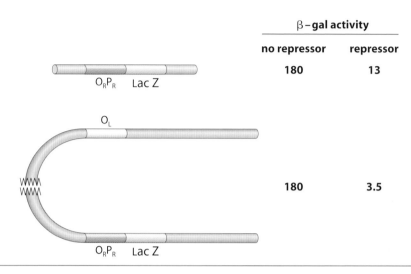

Figure 5.7. The helping effect of O_L on repression of P_R. The P_R-*lac* fusions used here are similar to that described in Fig. 4.19B, and repressor was supplied from a plasmid, also as described in that figure. The addition of the upstream sites increased repression under these conditions from 14-fold to 50-fold. The repressor concentration is probably relatively low in these experiments. When a full lysogen's worth of repressor is provided, the addition of O_L increases repression of P_R from some 80-fold to at least 500-fold (K. Shearwin, pers. comm.).

- O_L, positioned 3.6 kb from O_R, improves repression of P_R in bacteria (Fig. 5.7). The explanation for this effect is that repressor is helped to bind to O_R by interacting with repressor binding to O_L.

These experiments were inspired by (or at any rate followed by several years) the finding that, at high concentrations, purified repressor forms an octamer. We return to this important matter at the end of this section.

Reference: 50.

Long-range Interactions and Repression of P_{RM}

A lysogen induces poorly if the prophage bears a mutation in either O_R3 or O_L3 that prevents repressor binding to that site. Either mutation causes the repressor concentration to increase some 2.5- to 3-fold.

These results argue that repressor down-regulates transcription of its own gene in a lysogen; that even this 2.5- to 3-fold effect is crucial for the lysogen to be fully inducible; and that autonegative control depends on O_L3 as well as O_R3.

Reference: 22.

Activation and Repression of P_{RM}

The response of P_{RM} to a range of repressor concentrations was reexamined in an experiment similar to that of Figure 4.19A, but with a crucial addition: O_L was added 3.8-kb downstream from P_{RM}. As in the earlier experiment, repressor activates, and at higher concentrations represses, P_{RM}. The experimental results closely match the curves of Figure 5.5.

There is a limit, not well defined at the moment, to the distance over which repressor binding to one operator can influence repressor binding to another. Thus, O_L placed more than 500 kb away has no effect on the activity of P_{RM}.

References: 23, Ian Dodd, pers. comm.

Repressor Structure

In Chapter 4 (page 76) it was noted that purified repressor forms tetramers at concentrations well above that found in a lysogen. A more sophisticated sedimentation analysis reveals the formation of octamers as well (see the panel *Detection of the Repressor Octamer*). A convergence of crystallographic and genetic experiments has now provided us with a picture of the octamer, as follows.

Crystallography. Recall our old picture of repressor: a protein of two folded domains connected by a protease-sensitive linker (Fig. 1.6). The amino domain binds DNA (see the panel *Protein-DNA Interactions*), and the carboxyl domain dimerizes and forms higher-order oligomers (Fig. 1.7). The only crystallographic analysis supporting this picture was the structure of the amino domain separated from the rest of the protein.

DETECTION OF THE REPRESSOR OCTAMER

The surmise that repressor forms tetramers (but not higher-order species) was based primarily on analysis of its rate of sedimentation at high (versus low) concentrations (see page 76). Such experiments can be misleading: A given sedimentation profile can reflect the presence of a stable tetramer or, for example, it might represent a mixture of octamers and dimers in equilibrium with each other.

In a more sophisticated analysis, repressor was sedimented to equilibrium. At equilibrium, sedimentation and diffusion balance, and a smooth concentration gradient of repressor is produced. The shape of that gradient is determined by the relative amounts of the various oligomeric species present. Pure octamer, for example, would produce a steeper concentration gradient than would pure dimer. By repeating the experiment at many different repressor concentrations, and analyzing the data with a computer program, it was deduced that as the repressor concentration increases, repressor forms dimers, then tetramers, and those tetramers very rapidly associate to octamers. No trimers, hexamers, septamers, or species larger than the octamer were detected.

The biological significance of the octamer was not apparent for some years. One reason for this delay is that, on the assumption that octamers form in vivo with the same efficiency as in vitro, the octamer was predicted to form at very low levels in cells (well under one octamer per cell), a matter discussed in the text and in the panel *Repressor Oligomerization On and Off DNA*.

References: 51, 52, 48.

PROTEIN-DNA INTERACTIONS

As described in ChapterTwo, model building indicated that repressor and Cro use a common motif, called the helix-turn-helix motif, to dock with DNA. This surmise was later confirmed by the crystallographic structures of protein-DNA complexes. These structures have also largely confirmed the specific amino acid-base pair contacts made by repressor and Cro upon binding DNA, as set forth in Figure 2.11.

On page 42 two questions were raised: does the recognition helix present itself identically to the DNA in every case so that amino acids at specified positions interact with bases at specified positions in the operator and, is there a simple code describing amino acid-base pair interactions? Since those questions were raised, several additional helix-turn-helix proteins have been studied, and the answer to both questions is no.

References: 4, 12, 13, 17.

We now have the crystallographic structure of the repressor's carboxyl domain. This domain crystallizes as an octamer, and the structure shows how dimers, tetramers, and octamers form.

As diagrammed in Figure 5.8, three different surfaces mediate protein-protein interactions. One patch (colored white in the figure) directs dimerization: this patch on one monomer interacts with the corresponding patch on another to form the dimer as shown. Two additional patches (light and dark blue) mediate the dimer-dimer interactions responsible for tetramerization and octamerization.

The precise way the dimers fit together (difficult to draw accurately in two dimensions) ensures that repressor does not make an endlessly repeating polymer. Rather, four dimers form a closed circle, and no higher-order oligomers are formed. In the octamer, the DNA-binding domains project outward, positioned so as to bind the operators.

Figure 5.9 shows the repressor octamer positioned to interact with four operator sites.

References: 18, 19.

Genetics. Genetic experiments have identified residues that are required for repressor dimerization, on the one hand, and for cooperative binding of repressor dimers to DNA, on the other.

Mutants deficient in dimerization have been isolated by exploiting the following phenotype. Lysogens that express dimerization-deficient mutant repressors induce at unusually low doses of UV irradiation. This is because monomers, but not dimers, are cleaved upon induction, and thus poorly dimerizing repressors are more readily eliminated from the cell. Mutations that are confirmed to decrease dimerization, as assayed in vitro, map to the dimerization surface, as diagrammed in Figure 5.10.

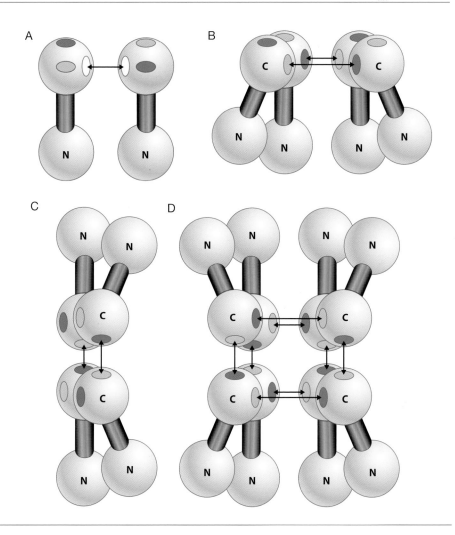

Figure 5.8. Formation of dimers, tetramers, and octamers. (*A*) Dimer formation is an example of an isologous interaction—i.e., identical surfaces (colored white) come together. This reaction stops once dimers are formed. Of the three interactions shown in this figure, this is the strongest. (*B*) The interactions to form tetramer are "heterologous"—a light blue patch pairs with a dark blue one. Note also that in the tetramer each of the two carboxyl domains of one dimer contacts a carboxyl domain in the other dimer. Thus figures in the text (Fig. 1.19, for example) that show "side-by-side" contact of two repressor dimers—in which only two carboxyl domains participate in tetramer formation—are oversimplifications. (*C*) An alternative way to form a tetramer. (*D*) An octamer is formed from two tetramers by apposition of the same kinds of surfaces as were used to form tetramers from dimers. The dimers bind at an angle to each other such that the reaction forms a closed octamer. Each patch is paired with a partner and buried.

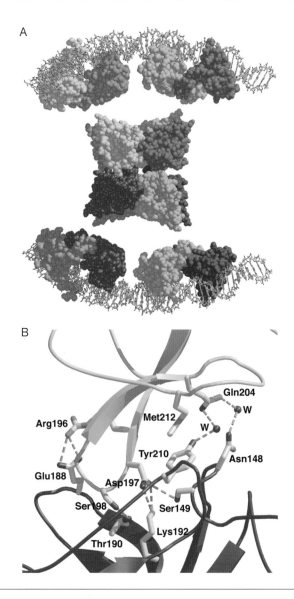

Figure 5.9. (*A*) The repressor octamer and DNA. The structure in the center is the octamer of eight carboxyl domains. The amino domains are bound to the DNA. In the actual structure the amino and carboxyl domains are contiguous. (*B*) The octamer interface. The green and red protein segments correspond to the light and dark blue patches of Figures 5.8 and 5.10. Note the extensive correspondence between the genetic and structural analyses: Each of the residues labeled here, with one exception, was found changed in one of the cooperativity mutants isolated as described in Fig. 5.12. Three additional sites of cooperativity mutations (Phe202, Gly147, and Gly199) lie on or near the depicted surface. The interactions of Gln204 with its partners (Tyr210 and Asn148) are mediated by water molecules as shown. (*References:* Beamer and Pabo 1992; Bell and Lewis 2001).) (Image prepared with BobScript, MolScript, and Raster3D.)

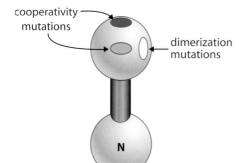

cooperativity mutations

dimerization mutations

N

Figure 5.10. Clusters of mutations that diminish repressor dimerization, in one case, and tetramerization in the other. The patches correspond to those in Fig. 5.8.

Mutants that can form dimers but not higher-order oligomers bind to DNA, but the DNA-bound dimers cannot interact with each other. The screen used to isolate such mutants exploited the following findings.

- We are familiar with the idea that two repressor dimers bind cooperatively to DNA fragments bearing two adjacent repressor-binding sites—as at O_R, for example. We now know that, as shown in Figure 5.11, two repressor dimers bind cooperatively to two separated operator sites, an effect observed with sites separated by up to about 100 base pairs. This cooperative binding differs from that seen when the two sites are adjacent (as with O_R1 and O_R2 in O_R), in the following way.

- The upstream repressor dimer prevents the downstream dimer from activating transcription (see bottom part of Fig. 5.12). Apparently the protein, the DNA, or both are distorted so that the amino domain adjacent to the promoter cannot appropriately interact with its target surface on polymerase.

Mutants of repressor that activate P_{RM} in the construct of the bottom part of Figure 5.12 have the following properties: they dimerize and bind DNA but lack the cooperativity function. Thus the mutant dimer bound upstream cannot interact with the downstream dimer, and so that downstream dimer is free to contact polymerase and thereby activate transcription. (In this experiment, sufficient repressor is supplied so that binding of repressor to O_R2 does not depend on interaction with the upstream repressor.)

The positions of the amino acid changes in these mutants map to two patches on the surface of the carboxyl domain as shown in Figure 5.10. These patches, according to the crystal structure, mediate tetramerization and octamerization.

These genetic results help interpret the crystallographic studies. The interactions that drive octamer formation in the crystals, as shown in Figure 5.9, involve a few ion pairs and hydrogen bonds. Without the genetic evidence, one might worry that these interactions are artifacts caused by biologically irrelevant crystal-packing forces.

References: 34, 33, 28, 55.

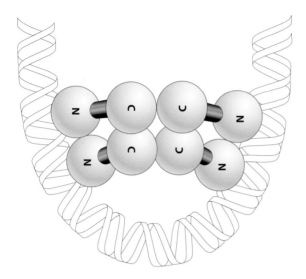

Figure 5.11. Binding of two repressor dimers to separated DNA sites. Two repressor dimers are shown bound to sites separated by 60 base pairs, six turns of the DNA helix. Repressor binds cooperatively to these sites just as it does when the sites are adjacent. That is, the affinity of the weak site is increased by the presence of an upstream site. The DNA must distort to make such a short loop. Over these short distances, the repressors must bind to the same side of the DNA helix to interact.

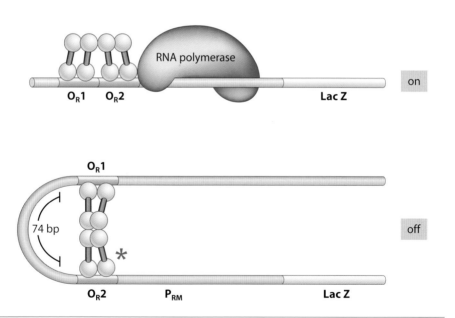

Figure 5.12. Inhibition of activation. In the reporter on the bottom, O_R1 has been moved 74 base pairs upstream of its usual position (as in the reporter on top), and now interaction of repressor dimers with each other hinders activation. Repressor mutants that have lost that interaction bind O_R2 and stimulate transcription.

Cro-binding to $O_R 3$ and Induction. On page 95 it was claimed that, following UV irradiation, the key event triggering induction was the binding of Cro to $O_R 3$. The basis for the claim was the finding that lysogens containing prophages bearing mutations in $O_R 3$ fail to induce when irradiated. This interpretation seemed inevitable according to our (mistaken) view that repressor did not significantly bind $O_R 3$ in a lysogen, and so the only relevant effect of the $O_R 3$ mutations would be to prevent Cro from binding to that site.

Now we know that overproduction of repressor in a lysogen—a consequence of mutating $O_R 3$—is sufficient to impair induction. So for now the conclusion that Cro must bind $O_R 3$ to trigger the transition to lytic growth, although not excluded, remains uncertain.

Reference: 22.

REPRESSOR OLIGOMERIZATION ON AND OFF DNA

We noted in the panel *Detection of the Repressor Octamer* that purified repressor forms octamers only at a concentration significantly higher than that found in a lysogen. In fact, as noted, these studies suggest that there is no more than about one free octamer for every ten cells, and perhaps about one or two free tetramers per cell. Nevertheless, the octamer forms as repressor binds simultaneously (and cooperatively) to O_L and O_R in a lysogen.

What forces drive this oligomerization on DNA? In thinking about this problem, it is helpful to first consider formation of the tetramer on DNA, and then consider interaction of two DNA-bound tetramers to form the DNA-bound octamer.

Formation of Repressor Tetramers on DNA

Repressors binding to adjacent sites (as at $O_R 1$ and $O_R 2$) are in very high local concentration with respect to one another, and so an interaction (in this case tetramerization) can occur on the DNA that does not occur off the DNA (where the effective concentration of repressor is much lower). It is estimated, however, that once these two tetramers are formed on DNA, an additional increase of some 500-fold in their effective concentration (compared with the concentration of two tetramers free in the cell) would be required for the observed octamer formation with DNA looping. Where does this additional concentration come from?

Interaction of DNA-bound Tetramers to Form the Octamer

Repressor tetramers bound to sites separated by 2–3 kb (as in a lysogen) are expected to be at higher concentration with respect to each other than are two tetramers free in the cytoplasm. This expectation follows from the number of base pairs separating the operators and the assumption that DNA behaves as a flexible linear polymer in cells. The effect is calculated (very approximately) to be tenfold.

But DNA in cells is likely to be even more compact than is deduced from the calculation just alluded to, thanks to the phenomenon of DNA supercoiling. A DNA supercoil (a negative supercoil in this case) results when closed circular DNA (such as the bac-terial chromosome) is opened, partially unwound, and closed again. The interwound DNA segments "slither" across one another and rapidly bring even widely separated sites into close apposition. This effect of DNA supercoiling, difficult to quantitate precisely, may be quite large.

Still another factor that can increase the interaction of two DNA-bound tetramers—compared to two tetramers free in solution—is that the DNA-bound repressors are pre-sented in relatively fixed orientations. Thus, some of the entropic factors that work against interaction between free repressors are eliminated by the prior formation of fixed tetramers on DNA.

Whatever the explanation might be for how the octamer forms on DNA, the reader should not lose sight of the essential finding: a molecular species that forms at extreme-ly low efficiency (when measured with purified protein) can nevertheless play an impor-tant role in gene regulation, in this case simultaneously recognizing sites separated by at least several thousand base pairs.

References: 23, 8, 6, 9, 54, 45.

---

2. POSITIVE CONTROL (ACTIVATION OF TRANSCRIPTION)

In Chapter One we noted that repressor works as an activator by touching RNA polymerase: The repressor bound at O_R2 contacts polymerase binding to the adja-cent promoter, P_{RM}. The text also identified the region of repressor that contacts polymerase. This "activating region," as we now call it, lies along the outside of repressor's helix 2 (Figs. 1.12, 2.17, and 4.28).

We now have a high-resolution picture of the repressor-polymerase interac-tion, and we can explain how that interaction stimulates transcription. Before sum-marizing these findings, we need to describe *E. coli* RNA polymerase and its inter-action with a typical promoter in greater detail than we did in Chapter One.

Polymerase and Promoter

Figure 5.13 illustrates the four essential subunits of the major form of *E. coli* RNA polymerase. One subunit, called σ70, recognizes most promoters in *E. coli*, includ-ing the λ phage promoters. One part of σ70 (its so-called region 4) recognizes the characteristic " –35 region" in the promoter, and another part of this elongated pro-tein recognizes the "–10 region." These two promoter regions are shown in Figure 2.15.

Polymerase binds a promoter in steps. It loosely associates with double-strand-ed DNA to form the so-called "closed" complex. This complex spontaneously iso-merizes to the so-called "open" complex in which ten base pairs around the start site of transcription are separated to expose the template strand. Formation of the open complex at most promoters is essentially irreversible, and transcription quickly ensues in the presence of substrate (the nucleoside triphosphates).

Recall that at "strong" promoters, such as P_R and P_L, polymerase efficiently binds and transcribes on its own. But at weak promoters (which bear −35 and/or −10 sequences that differ slightly from the canonical ones), efficient transcription requires an activator. P_{RM} is a weak promoter that is helped to work more efficiently by interaction of λ repressor with polymerase.

Reference: 10.

The Mechanism of Activation

We now know that λ repressor's activating region contacts the σ70 subunit of polymerase. Only a few contacts, distributed over a small area, are involved. The interaction is weak, no obvious change in shape of either protein accompanies the interaction, and no hydrolysis of ATP (or of any other such energy source) is involved.

Activation of transcription by λ repressor thus resembles the examples of cooperative binding we have discussed, but with an added feature: one of the partners—RNA polymerase—forms an irreversible (open) complex with DNA. Repressor stabilizes polymerase at the promoter until open complex formation is complete. In effect, as explained more extensively in the panel *Gene Activation in More Detail,* repressor "recruits" polymerase to the gene.

HOW DO WE KNOW

Mutations in repressor's activating region, and a mutation in σ, have been instrumental in describing the repressor-polymerase interaction that mediates activation of transcription.

Activating Region Variants

The *pc* (positive control) mutants of repressor that were first isolated, please recall, bind DNA but do not activate transcription (p. 100). Many additional activating region variants have been generated by changing amino acid residues along the outside of helix II. Although most of these work less well than wild type, some work as well as or even better than wild type (Figure 5.14).

As shown in the figure, all the improved activating regions retain Glu at position 2 of helix II, and most retain Asp at position 38. The results suggest that the two residues—Glu34 and Asp38—make crucial contacts with polymerases.

References: 29, 21, 32.

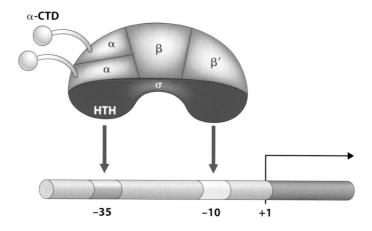

Figure 5.13. The four essential subunits of *E. coli* RNA polymerase and a typical promoter. The polymerase has two copies of α, a subunit comprising two domains connected by a linker. The cell produces several σ subunits, each of which directs the polymerase to a specific class of promoters. σ70, the one present in the polymerase shown here, is the predominant form. Region four of σ70 comprises the 88 amino acids at the carboxyl terminus of this 613-residue protein. This part of σ contains a helix-turn-helix motif that recognizes the –35 region of the promoter. The polmerase has two copies of α, a subunit comprising two domains connected by a linker. Some strong promoters have a sequence (called an "up element") to which the α carboxy-terminal domain (α-CTD) binds. A fifth subunit, ω, not shown here, is dispensable.

λ **repressor helix 2**

		inside			fold activation
	gln 33	val ala			
	glu – ser		asp – lys		7
		outside			

						fold activation
1	– glu	ser	–	– asp	gly	13
2	– glu	ser	–	– asp	glu	16
3	– glu	asp	–	– asp	glu	20
4	– glu	asp	–	– asp	glu	23
5	– glu	asp	–	– asp	asn	23
6	– glu	leu	–	– tyr	glu	23
7	– glu	leu	–	– tyr	asn	28

Figure 5.14. Activating region strengths of λ repressor variants. "Fold activation" denotes the measured increase in transcription from P_{RM} over the basal (unstimulated) level, elicited by each DNA-bound repressor variant. We say that, for example, variant 7's activating region is four times as strong as that of wild-type repressor.

A Suppressor of a *pc* Mutant

The target of λ repressor's activating region—σ—was originally defined as follows. Starting with a specific *pc* mutant repressor, a mutant polymerase was identified that (unlike wild-type polymerase) can be activated by the *pc* mutant. The repressor mutant bears a change at position 38 (Asp Asn) and the "suppressor" mutation in polymerase changes the residue at position 596 of σ with His. The results suggest that, in the wild-type case, the negatively charged Asp38 of repressor interacts with the positively charged Arg596 of σ. Position 596 lies within the part of σ (i.e., its region 4) that recognizes the –35 region of the promoter, and hence a part that lies close to repressor bound to DNA just upstream.

References: 41, 39.

Crystallography

We now have a crystallographic model of repressor (amino domains only) bound to DNA and contacting RNA polymerase bound to an adjacent promoter (Fig. 5.15). The structure shows that one repressor monomer contacts region 4 of σ over a small area involving only five residues, two on repressor and three on σ (Fig. 5.15B). The interaction causes no significant changes in the shapes of either protein.

The structure shows that, as predicted by the *pc* mutant-suppressor pair described above, repressor's Asp38 contacts σ's Arg596. Asp38 makes an additional contact (with σ's Lys593). Consistent with the results shown in Figure 5.14, repressor's Glu34 also contacts σ—its partner, revealed by the structure, is σ's Arg588.

In a preceding section (The Mechanism of Activation) it was asserted that activation of transcription by λ repressor resembles cooperative binding of repressors to DNA: a simple "adhesive" interaction recruits polymerase (in the activation case) to DNA. The results of genetic and crystallographic experiments described thus far are consistent with this view. The remainder of this section describes additional experiments that further support this picture.

References: 36, 47.

Activator Bypass

The following experiments show that evidently any binding reaction—protein-protein or protein-DNA—that stabilizes polymerase at P_{RM}, allowing it to form the open complex with DNA, suffices for activation. In these so-called "activator bypass" experiments, activation is effected in the absence of the ordinary activator-polymerase interaction.

- *Borrowed protein-protein interactions.* In the experiment of Figure 5.16, one protein is fused to a DNA-binding domain (and thereby tethered to DNA), and the other to polymerase itself. Transcription is activated if these two proteins interact.

- *A borrowed protein-DNA interaction.* In this experiment (Fig. 5.17) a protein

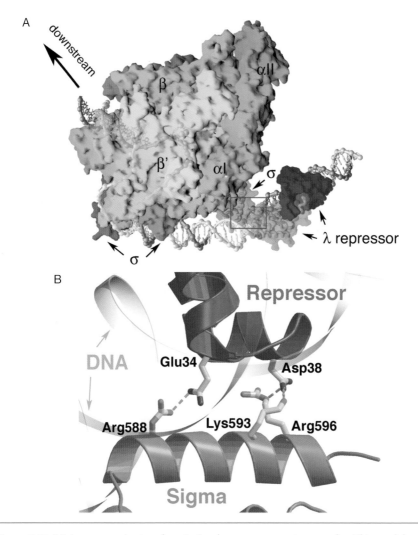

Figure 5.15. (*A*) A repressor (amino domains)-polymerase-promoter complex. This model was built by combining two protein-DNA crystallographic structures. The first had repressor (the amino domains) at the strong operator site O_L1 and a fragment of σ (region 4) bound to an adjacent –35 region of a promoter. The second is the complete RNA polymerase bound at a promoter. The polymerase and σ fragment were isolated from the bacterium *Thermus aquaticus*, and are very similar to the corresponding *E. coli* proteins. The loosely tethered α-CTD, diagrammed in the previous figure, is not seen in these structures. The ω subunit is hidden behind the α subunits. (*B*) The contacts between repressor and σ70 in the structure of part *A*. The five residues involved in forming this protein-protein interface cover only some 342 Å^2 of surface area. In addition to showing the critical contacts made by repressor's Glu34 and Asp38, the structure plausibly explains at least some of the substitutions of Fig. 5.14 that increase activating region strength. For example, a Glu in place of repressor's Lys39 would likely make an additional contact with σ's Arg596. Not shown here is an additional contact made by σ's Arg 588: Arg588 contacts σ's Glu585, positioning that residue so that it contacts DNA. (Reference: Jain D. et al. 2004 and D. Jain, pers. comm.) (*B*, Image prepared with BobScript, MolScript, and Raster3D.)

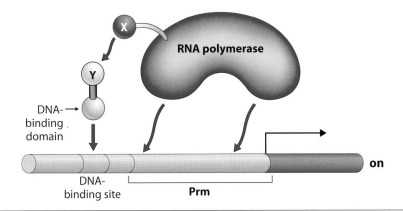

Figure 5.16. An activator bypass experiment. Protein X has been fused to polymerase (replacing the α-CTD) and protein Y has been fused to a DNA-binding protein (e.g., λ repressor). Interaction of X and Y recruits polymerase to the promoter and thereby activates transcription. Many different protein pairs—including a pair of yeast proteins—work in this experiment.

that binds DNA has been fused to a subunit of polymerase. The modified polymerase works at a high level at a gene bearing a binding site positioned just upstream of the promoter.
References: 25, 27, 24.

Changing Activating Regions and Target Context

Not only can heterologous interactions mediate gene activation as we have just seen; so can the ordinary activating-region–polymerase interactions work even

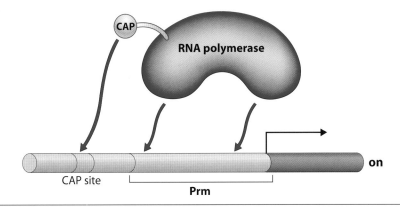

Figure 5.17. Another activator bypass experiment. The DNA-binding protein CAP has been fused to polymerase. The new form of polymerase initiates transcription efficiently if the promoter bears an added site upstream to which CAP binds. (CAP is itself an activator, but a *pc*-mutant CAP was used so that activation by the ordinary contact did not occur.) Other DNA-binding domains have also been shown to work in such experiments.

when their protein contexts are changed. This finding, illustrated in the following two experiments, reinforces our surmise that simple binding interactions mediate gene activation.

- If a surface of Cro is modified so as to resemble repressor's activating region, it can work as an activator of P_{RM}. The modified part of Cro is shown in Figure 5.18. The effect is seen when Cro is bound to O_R2 but not to O_R3.

- In the "caboose" experiment of Figure 5.19, σ's region 4 has been fused to the α subunit of polymerase in place of α's CTD. Now a repressor binding upstream of its ordinary position at P_{RM} can contact this added fragment and activate transcription. *References: 20, 26.*

Many different activators are found in *E. coli* that activate the polymerase bearing σ70. Each of these activators recognizes its own sites on DNA and contacts polymerase. Some of these activators contact the σ subunit others the α subunit. For example, the activator of the *lac* genes, called CAP, contacts the α-CTD when working at the *lac* promoter.

Thus, there are many different sites on polymerase that can be contacted by activators to stimulate transcription. It might be imagined that promoters differ in some subtle way such that stimulation of transcription at any given one requires some specified contact. But this is often not the case: on a DNA construct bearing a CAP-binding site positioned upstream of P_{RM}, CAP stimulates transcription from that promoter.

References: 3, 7

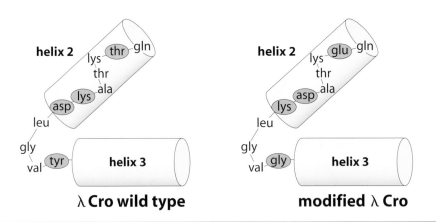

λ **Cro wild type**   **modified** λ **Cro**

Figure 5.18. Turning Cro into an activator. The modified Cro activates P_{RM} only if O_R3 has been heavily mutated so that Cro cannot bind there. The experiment works—i.e., the modified Cro activates P_{RM} on this template—because the disposition of Cro's helix-turn-helix (HTH) motif, when Cro is bound to DNA at O_R2, is very similar to that of repressor's HTH when repressor is bound at that site. The Asp-Lys change at the last position of helix 2 is required for efficient DNA binding. The other three circled positions correspond to sites of *pc* mutations as described in Fig. 4.28.

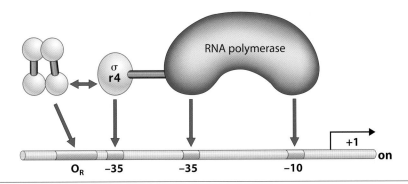

Figure 5.19. The "caboose" experiment. The polymerase bears its usual intact σ70 plus, as a caboose, an added fragment comprising σ region 4 (σr4). The promoter was modified so as to bear an extra –35 region and, just upstream of that element, a repressor-binding site. Repressor (provided at high concentration so that this single operator site is filled) stimulates transcription by contacting the added σ region 4. *pc* mutations that block ordinary activation block activation in this case as well.

The notion that the activator simply stabilizes polymerase at P_{RM} predicts that two contacts between activator and polymerase will work synergistically. That is, the two contacts working together will elicit more activation than the sum of the two interactions manifested separately. This general prediction is realized in the experiment of Figure 5.20. Here two activators—CAP and λ repressor—make simultaneous contacts with polymerase at P_{RM}. The effect of each of these activators can be measured in the absence of the other, and the two activators working together do so synergistically.
Reference: 37.

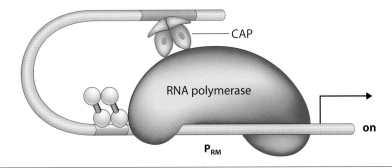

Figure 5.20. Synergistic activation. The DNA upstream of P_{RM} has been modified so as to bear a single strong λ repressor-binding site, and further upstream, a CAP binding site. When CAP alone works on this reporter, transcription is stimulated some tenfold. λ repressor, working alone, has a similar effect. But when the two activators work together, transcription is stimulated some 50-fold. In an ideal case, the binding energies add together. In such a case, the binding constant describing the affinity of the polymerase for the promoter increases exponentially as multiple activators work simultaneously. (CAP likely contacts the α-CTD here as it does when working at one of its natural sites found in the *lac* promoter.)

GENE ACTIVATION IN MORE DETAIL

Experiments described in the text suggest that gene activation by λ repressor is an example of cooperative binding of proteins to DNA. Thus, just as one λ repressor molecule can help another bind to its operator site on DNA, so can a repressor help RNA polymerase bind to its promoter site on DNA. In both cases a simple protein-protein interaction is required.

Beyond that simple description, however, we can specify more than one way that the reaction might proceed. Recall that the initial polymerase-promoter complex undergoes a spontaneous conformational change (i.e., the closed to open complex as described in an earlier part of this section).

The activator-polymerase interactions might stabilize the initial binding of polymerase to DNA (thereby stabilizing a closed complex), or the same kind of protein-protein interaction might stabilize a subsequent step leading to formation of the open complex. The net effect would be the same in both cases: the action of the activator would increase the rate of formation of stable (i.e., open) polymerase-promoter complexes.

Kinetic experiments suggest that CAP, working at the *lac* promoter, stabilizes the initial binding of polymerase. In contrast, wild-type λ repressor, interacting with wild-type polymerase at P_{RM}, stimulates a step beyond the initial binding of polymerase. This apparent difference between the action of CAP and λ repressor is readily altered. Thus, when working at P_{RM}, λ repressor stimulates the initial binding of a mutant polymerase. The polymerase mutant bears a single amino acid change and, other than the property just described, appears normal.

In each of the scenarios described above, in the absence of the activator, polymerase is detected only rarely at the promoter. In the presence of the activator, the polymerase quickly becomes stably bound (in the open complex with DNA). The general term used to describe this effect is "recruitment": We say that the activator recruits the polymerase to the promoter. Although, as we have noted, certain details of the reaction might differ, in each case a protein-protein interaction stabilizes polymerase on the DNA, and similar adhesive-like interactions between the activator and polymerase are required.

It is not hard for evolution to produce a peptide (an activating region) that stimulates transcription by recruitment when tethered to DNA. Recall that the polymerase-repressor interacting surfaces, like those that mediate cooperative binding of repressor to DNA, are small, involving just a few residues on each protein. The interactions are weak, so weak in fact, that they have not been observed with purified polymerase and repressor in the absence of DNA. On DNA, the polymerase and repressor-binding sites—P_{RM} and O_R2— are in very high concentration with respect to one another, and this helps drive the reaction. Not all activators in bacteria work by recruitment, and in those other cases, more highly articulared activator-polymerase interactions are required.

References: 30, 40, 19a, 7.

3. THE STRUCTURE OF THE REPRESSOR MONOMER AND THE MECHANISM OF REPRESSOR CLEAVAGE

Throughout the text λ repressor is drawn as a dumbbell: two domains, of similar size, are separated by an extended linker. This picture was based partly on the fact that the protease papain, when added to purified repressor, cuts the linker in several places to yield the two stable domains (Fig. 4.10). The result suggested that the linker (residues 93–132) was not tightly folded—hence its protease sensitivity—and so for convenience we drew it as an extended chain.

We now have a better picture of the repressor monomer, thanks initially to analysis of the crystal structure of its close relative LexA (Fig. 5.21). There are indeed two domains, but the linker, instead of being extended, lies along a surface of the carboxyl domain.

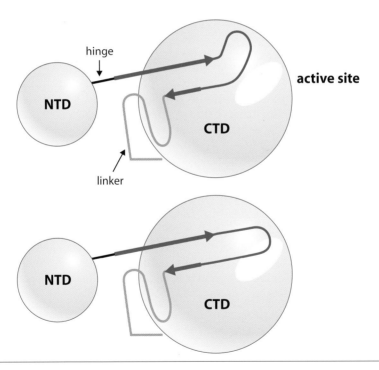

Figure 5.21. Two crystallographic forms of the LexA monomer. A hinge of four residues, rather than an extended linker, separates the two domains. In the structure on the bottom, the cleavage-sensitive Ala-Gly pair lies very near the catalytic Ser, and in the structure on the top it lies some 20 Å removed from that Ser. The structure of wild-type λ repressor is similar overall. The catalytic Ser is at position 149 in λ repressor and at position 119 in LexA. The cleavage-sensitive Ala-Gly is at postions 111–112 in λ repressor and at 84–85 in LexA repressor.

This refined structure of repressor helps explain the mechanism of repressor inactivation that follows exposure of lysogens to UV irradiation. The original text noted that repressor inactivation required the activity of the protein RecA, a protein ordinarily involved in DNA recombination. In Figure 1.21, RecA was pictured as a protease that, once activated by binding to damaged DNA, cleaves repressor at a single site in the linker.

But RecA is not at all a typical protease—in particular, it does not have an active site that cleaves repressor. Rather, the active site, which includes a catalytic Ser residue as do many proteases, lies on the surface of the carboxyl domain of repressor itself. Thus, repressor cleaves itself, an autocatalytic reaction that, in the cell, proceeds slowly in the absence of activated RecA. RecA evidently stabilizes a conformation of the linker that juxtaposes the cleavage site with the protease active site on the carboxyl domain.

HOW DO WE KNOW

Both λ repressor and its close relative LexA spontaneously cleave when purified and held at high pH. The rates of these reactions, slow at neutral pH, are increased by addition of RecA. In both the self- and RecA-mediated cleavage reactions, a specific Ala-Gly bond in the linker (in both λ repressor and LexA) is broken.

Mutant repressors that cleave more readily than wild type, as well as other mutants that cleave less well than wild type, have been isolated.

- A mutant repressor that is not cleaved in vivo upon induction also fails to cleave when held at high pH (whether or not RecA is present). An example of such a mutant is one bearing a change at the site of a crucial Ser residue in the carboxyl domain.

- A mutant of the opposite type—one that cleaves abnormally quickly in the cell upon induction—also cleaves more quickly when held at high pH. An example of such a mutant is one that dimerizes poorly.

The relationship between the self- and RecA-mediated cleavage reactions emerges from analysis of crystal structures, particularly those of LexA and certain of its mutant variants. The models shown in Figure 5.21 illustrate several important findings of these experiments.

- In the intact monomer, the linker is mostly folded onto the carboxyl domain. The region separating the domains (four residues) resembles a hinge more than an extended linker.

- The protease active site on the carboxyl domain—which includes a crucial Ser residue as shown—lies at the end of a cleft. The linker fits into this cleft in one of the structures, with the sensitive Ala-Gly pair positioned close to the active site. In another structure, the Ala-Gly pair is positioned well away from the active site.

- Mutations that increase or decrease the rate of repressor cleavage (in addition to those on the dimerization surface) are found in and around the cleavage-sensitive Ala-Gly, and also in and around the cleft that binds this region of the linker.

Evidently, then, the LexA linker can adopt two conformations, in one of which its sensitive Ala-Gly bond lies close to the protease active site on the carboxyl domain. RecA evidently binds to repressor and stabilizes this positioning of the linker. It is plausible that this function of RecA is mimicked by increasing the pH.

The crystallographic structure of an intact λ repressor monomer (in this case as part of a dimer bound to DNA) resembles that of the LexA, with most of the linker folded onto the carboxyl domain. It is thus likely that the conclusions drawn from the LexA studies described here apply as well to λ repressor.

References: 42, 43, 46, 28.

4. EVOLVING THE SWITCH

The switch appears to us today in a sophisticated and intricate form, one that depends on a series of carefully modulated protein-protein and protein-DNA interactions. How might this set of interactions have evolved?

Keep in mind the overall biological function of the switch: the phage must propagate in two modes (in a lysogen and lytically), and an extracellular signal (UV irradiation) must induce a change from lysogenic to lytic growth. In its current form, the switch is highly efficient: cells spontaneously induce very rarely, but virtually every cell induces in response to an appropriate signal.

One way to investigate how the switch might have assembled is to partially disassemble it—that is, to modify or eliminate certain features, and see how the switch is affected. Preliminary forays of this sort suggest that the switch still works, but less efficiently, when subjected to several kinds of changes. That surmise, in turn, encourages us to imagine that the switch evolved stepwise, with each added feature improving its function.

Here are three kinds of mutations that affect operation of the switch. The first changes the affinities of sites in O_R for repressor, the second elminates positive control, and the third eliminates cooperativity between DNA-binding dimers.

Changing the Affinities of Sites in O_R for Repressor

In one experiment, the sequence of $O_R 1$ was changed to make it identical to that of $O_R 3$. This mutation changes, as expected, the pattern of repressor binding to O_R. Now, as studied with a DNA fragment bearing O_R, repressor binds cooperatively, and at the same concentration, either to $O_R 1$ and $O_R 2$, or (on an equal fraction of DNA molecules), to $O_R 2$ and $O_R 3$.

A phage bearing this mutant operator (called 323) is deficient in several ways compared to wild type: it lysogenizes cells about half as efficiently; the rate of spontaneous induction of the lysogens is increased some 100- to 1000-fold; lytic growth produces about half as many new phage per growth cycle; and the lysogens induce at a lower dosage of UV light.

Starting with 323 (which differs from wild type by three changes), better-growing mutants bearing additional changes have been isolated. In one of these variants, a mutant residue in 323 has been changed back to its wild-type counterpart. This variant has thus evolved toward wild type under selective pressure for better lytic growth.

Eliminating Positive Control

Positive control—transcriptional activation of its own gene by repressor—has been eliminated in two ways. In one case a *pc* mutation was introduced into repressor expressed by an otherwise wild-type phage. In another, a mutation that prevents activation by λ repressor was introduced into the bacterium's sigma gene.

The mutant phage formed lysogens in a wild-type cell, and a wild-type phage formed lysogens in the mutant cell. But, in both cases these lysogens were aberrant: they lysed spontaneously at an unusually high rate, and they were induced by unusually low levels of UV light. These properties are as expected if the repressor levels in the mutant lysogens are lower than in wild-type lysogens.

Eliminating Cooperativity between DNA-binding Dimers

A mutation that prevents repressor tetramerization and octamerization was introduced into a phage—that phage cannot lysogenize. When combined with a mutation in P_{RM} that makes that promoter work a bit better on its own, together with a mutation in O_R2 that makes repressor bind a bit tighter there, the triply mutant phage lysogenizes and grows lytically. The lysogens induce in response to UV irradiation, but the yield of new phages is reduced compared to wild type.

The general conclusion from these early probings is that the switch need not have been selected in some single, concerted, step. Rather, there are plausible

---

Figure 5.22. Hypothetical steps in the evolution of the switch. Stage 1. On this hypothetical primitive λ genome P_R and P_{RM} are well separated, and the single indicated operator site overlaps P_R. Repressor bound to that site would turn off lytic genes but have no effect on P_{RM}. (An additional operator site would control P_L). Perhaps such a phage could grow both lytically and form lysogens. Stage 2. The single depicted operator site has been moved close to a weakened P_{RM}, and now, by virtue of a contact between polymerase and DNA, synthesis of repressor is autocatalytic. A lysogen of such a phage would, presumably, induce more efficiently because repressor synthesis would decrease as repressor was inactivated. Stage 3. A second operator site has been introduced, and repressor (by virtue of a protein-protein contact) binds cooperatively to these two sites. The affinities of the sites are adjusted so that, at the concentration of repressor in a lysogen, binding of repressor to O_R2 depends on "help" from a repressor binding to O_R1. This improvement stabilizes the lysogens: the newly introduced cooperativity ensures that small, spontaneous decreases in repressor concentration will not lead to induction. Rather, the lysogens respond in an all-or-none fashion when the repressor reaches a critical level. Stage 4. The position of the third repressor binding site (O_R3) allows repressor to negatively regulate its own synthesis. The repressor concentration never exceeds a specified level, a stipulation that helps ensure an efficient switching mechanism. (Redrawn, with permission, from Ptashne, M. and Gann, A. 1998. Curr. Biol. 8, R812–822, p. R818. ©1998 Elsevier Science Ltd. All rights reserved.)

intermediates that did not work as well as the presently constituted switch. Thus, natural selection can develop a switch—within the narrow confines of the current discussion—stepwise, each step producing a more efficient version.

Reference: 44.

A purely hypothetical scheme for how repressor action at O_R might have evolved is shown in Figure 5.22. According to this "just so" story (as evolutionists

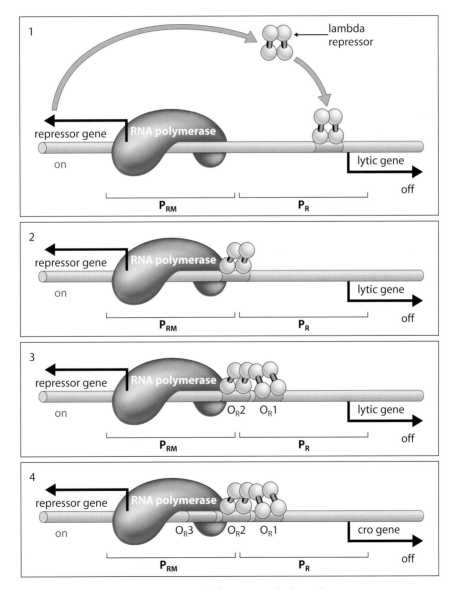

Figure 5.22. (*See facing page for legend.*)

would call it), the earliest proposed form has no cooperativity and no positive control. The second introduces positive control, the third cooperativity, and the fourth autogenous negative control.

Our understanding of the reactions underlying each of these "improvements" suggests that each step is not particularly difficult to take.

- Recall that both positive control and cooperativity require interacting surfaces involving just a few amino acids. The proteins need not change their shapes in any special ways, and there are only limited stereospecific constraints on these reactions. The very simplicity of the underlying mechanisms—simple binding interactions—would seem to make these changes "easy."

- Surface changes of another kind are also introduced at each step: moving, or introducing, a repressor-binding site on DNA. The affinities of those sites for repressor can then be properly modified by minimal changes in operator sequence.

5. CII AND THE DECISION

As described in Chapter One, bacterial cells infected with λ phages face a decision: whether to support lytic growth of the phage or to form lysogens. The state of the "switch" reflects the decision: a phage on the lytic pathway expresses predominantly Cro but not repressor, and vice versa for a phage destined to lysogenize the cell. In Chapter Three (p. 56) we noted that our understanding of how the decision is made is rudimentary, beyond the important surmise that the concentration of the protease-sensitive protein CII plays a key role.

We have since learned more about the action of CII. Not only does this protein activate transcription of the repressor gene (from P_{RE}) and of the *int* genes (from P_{int}) as shown in Figure 3.9; as shown in Figure 5.23, CII also activates transcription of "anti-sense" Q RNA from the promoter P_{AQ} (anti-Q). The "anti-Q" RNA pairs with the Q mRNA, and the resultant double-stranded RNA is destroyed by the enzyme (RNase III). Activation of P_{AQ} by CII is important for lysogeny: mutants with damaged P_{AQ} follow the lytic pathway much more frequently than does wild type. Thus, a phage destined to lysogenize the cell, thanks to CII, activates genes required for lysogeny—*cI* and *int/xis*—as it inhibits expression of a gene required for lytic growth—*Q*.

Our understanding of the "Decision" has not advanced much since the appearance of the original *Genetic Switch*. We know (and knew then) that environmental factors can have a strong influence on the fraction of infected cells that follow one or the other pathway presumably. But why, under any given set of conditions, do some infected cells lyse and some lysogenize? Part of the answer lies in the distribution of phages among the bacteria: those bacteria infected by multiple phages (as opposed to one or a few) are more likely to become lysogens. It seems, however, that cells infected with the same number of phages under identical environ-

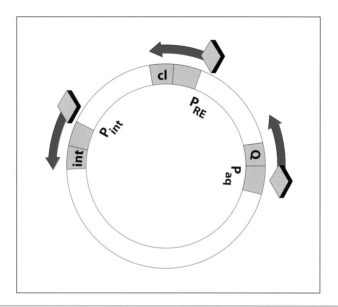

Figure 5.23. A third site of action of CII: the "anti-Q" promoter. This figure is identical to Fig. 3.9 with the addition of a third site of action of CII: the so-called anti-Q promoter (P_{AQ}). CII is believed to work at P_{AQ} just as it does at P_{RE} and at P_{int}: binding to a CII site, it recruits polymerase to the adjacent promoter.

mental circumstances sometimes lyse and sometimes lysogenize. What chance events, occurring before or after infection, determine the outcome remain to be elucidated.

References: 1, 2, 5, 31, 35, 38.

Attempts have been made to mathematically model various aspects of the switch—the stability of lysogens, the lysis-lysogeny decision, and so on. References to some of these papers follow.

References: 11, 14, 15, 16, 49, 53.

REFERENCES: BOOKS AND REVIEWS

1. Campbell, A. (2003). The future of bacteriophage biology. *Nat. Rev. Genet.* 4, 471–477.

2. Friedman, D.I. and Court, D.L. (2001). Bacteriophage lambda: Alive and well and still doing its thing. *Curr. Opin. Microbiol.* 4, 201–207.

3. Hochschild, A. and Dove, S.L. (1998). Protein-protein contacts that activate and repress prokaryotic transcription. *Cell* 5, 597–600.

4. Pabo, C.O. and Nekludova, N. (2000). Why is there no simple code for recognition? *J. Mol. Biol.* 301, 597–624.

5. Nudler, E. and Gottesman, M.E. (2002). Transcription termination and anti-termination in *E. coli. Genes Cells* 7, 755–768.

6. Ptashne, M. (1986). Gene regulation by proteins acting nearby and at a distance. *Nature* 332, 697–701.

7. Ptashne, M. and Gann, A. (2002). *Genes and signals.* (New York: Cold Spring Harbor Laboratory Press, Cold Spring Harbor.)

8. Rippe, K. (2001). Making contacts on a nucleic acid polymer. *Trends Biochem. Sci.* 26, 733–740.

9. Stark, M.W. and Boocock, M.R (1995). Topological selectivity in site-specific recombination. In *Mobile genetic elements,* D. Sherratt, ed. p. 179. (Oxford: IRL Press).

10. Young, B.A., Gruber, T.M., and Gross, C.A. (2002). Views of transcription initiation. *Cell* 109, 417–420.

REFERENCES: RESEARCH ARTICLES

11. Ackers, G.K., Johnson, A.D., and Shea, M.A. (1982) Quantitative model for gene regulation by a lambda phage repressor. *Proc. Natl. Acad. Sci. USA* 79, 1129–1133.

12. Albright, R.A. and Matthews, B.W. (1998). How Cro and λ repressor distinguish between operators: The structural basis underlying a genetic switch. *Proc. Natl. Acad. Sci. USA* 95, 3431–3436.

13. Albright, R.A. and Matthews, B. (1996). Crystal structure of lambda-Cro bound to a consensus operator at 3.0 Å resolution. *J. Mol. Biol.* 280, 137–151.

14. Arkin, A., Ross, J., and McAdams, H.H. (1998). Stochastic kinetic analysis of development pathway bifurcation in phage lambda-infected *Escherichia coli* cells. *Genetics* 149, 1633–1648.

15. Aurell, E., Brown, S., Johanson, J., and Sneppen, K. (2002) Stability puzzle in phage λ. *Phys. Rev. E* 65, 051914–4307.

16. Aurell, E. and Sneppen, K. (2002) Epigenetics as a first exit problem. *Phys. Rev. Lett.* 88, 048101-1-4.

17. Beamer, L.J. and Pabo, C.O. (1992). Refined 1.8 Å crystal structure of the lambda repressor-operator complex. *J. Mol. Biol.* 1, 177–196.

18. Bell, C.E., Frescura, P., Hochschild, A., and Lewis, M. (2000). Crystal structure of the λ repressor C-terminal domain provides a model for cooperative operator binding. *Cell* 101, 801–811.

19. Bell, C.E. and Lewis, M. (2001). Crystal structure of the λ repressor of C-terminal domain octamer. *J. Mol. Biol.* 314, 1127–1136.

19a. Benoff, B., Yang, H., Lawson, C.L., Parkinson, G., Liu, J., Blatter, E., Ebright, W., Berman, H.M., and Ebright, R.H. 2002. Structural basis of transcription activation: The CAP-CTD-DNA complex. *Science* 297, 1562–1566.

20. Bushman, F.D. and Ptashne, M. (1988). Turning λ Cro into a transcriptional activator. *Cell* 54, 191–197.

21. Bushman, F.D., Shang, C., and Ptashne, M. (1989). One glutamic acid residue plays a key

role in the activation function of lambda repressor. *Cell* 58, 1163–1171.

22. Dodd, I.B., Perkins, A.J., Tsemitsidis, D., and Egan, J.B. (2001). Octamerization of a λCI repressor is needed for effective repression of a Prm and efficient switching from lysogeny. *Genes Dev.* 15, 3013–3022.

23. Dodd, I.B., Shearwin, K.E., Perkins, A.J., Burr, T., Hochschild, A., and Egan, J.B. (2004). Cooperativity in long-range gene regulation by the lambda CI repressor. *Genes Dev. 18* (in press).

24. Dove, S.L. and Hochschild, A. (1998). Conversion of the omega subunit of *Escherichia coli* RNA polymerase into a transcriptional activator or an activator target. *Genes Dev.* 12, 745–754.

25. Dove, S.L. and Hochschild, A. (1998). Use of artificial activators to define a role for protein-protein and protein-DNA contacts in transcriptional activation. *Cold Spring Harb. Symp. Quant. Biol.* 63, 173–180.

26. Dove, S.L., Huang, F.W., and Hochschild, A. (2000). Mechanism for a transcriptional activator that works at the isomerization step. *Proc. Natl. Acad. Sci. USA* 97, 13215–13220.

27. Dove, S.L., Joung, J.K., and Hochschild, A. (1997). Activation of prokaryotic transcription through arbitrary protein-protein contacts. *Nature* 386, 627–630.

28. Gimble, F.S. and Sauer, R.T. (1989). Lambda repressor mutants that are better substrates for RecA-mediated cleavage. *J. Mol. Biol.* 206, 29–39.

29. Guarente, L., Nye, J.S., Hochschild, A., and Ptashne, M. (1982). Mutant lambda phage repressor with a specific defect in its positive control function. *Proc. Natl. Acad. Sci. USA* 79, 2236–2239.

30. Hawley, D.K. and McClure, W.R. (1982). Mechanism of activation of transcription initiation from the lambda PRM promoter. *J. Mol. Biol.* 3, 493–525.

31. Ho, Y.S. and Rosenberg, M. (1985). Characterization of a third, cII-dependent, coordinately activated promoter on phage lambda involved in lysogenic development. *J. Biol. Chem.* 260, 11838–11844.

32. Hochschild, A., Irwin, N., and Ptashne, M. (1983). Repressor structure and the mechanism of positive control. *Cell* 32, 319-25.

33. Hochschild, A. and Ptashne, M. (1986). Cooperative binding of lambda repressors to sites separated by integral turns of the DNA helix. *Cell* 44, 681–668.

34. Hochschild, A. and Ptashne, M. (1988). Interaction at a distance between lambda repressors disrupts gene activation. *Nature* 336, 353–357.

35. Hoopes, B.C. and McClure, W.R. (1985). A cII-dependent promoter is located within the Q gene of bacteriophage lambda. *Proc. Natl. Acad. Sci. USA* 82, 3134–3138.

36. Jain, D., Nickels, B.E, Sun, L., Hochschild, A., and Darst, S. (2004). Structure of a ternary transcription activation complex. *Mol. Cell 13,* 45–53.

37. Joung, J.K., Koepp, D.M., and Hochschild, A. (1994). Synergistic activation of transcription by bacteriophage lambda cI protein and *E. coli* cAMP receptor protein. *Science* 265, 1863–1866.

38. Kobiler, O., Koby, S., Teff, D., Court, D., and Oppenheim, A.B. (2002). The phage λ CII transcriptional activator carries a C-terminal domain signaling for rapid proteolysis. *Proc. Natl. Acad. Sci. USA* 99, 14964–14969.

39. Kuldell, N. and Hochschild, A. (1994). Amino acid substitutions in the −35 recognition motif of sigma 70 that result in defects in phage lambda repressor-stimulated transcription. *J. Bacteriol.* 10, 2991–2998.

40. Li, M., McClure, W.R., and Susskind, M.M. (1997). Changing the mechanism of transcrip-

tional activation by phage lambda repressor. *Proc. Natl. Acad. Sci. USA* 94, 3691–3696.

41. Li, M., Moyle, H., and Susskind, M.M. (1994). Target of the transcriptional activation function of the phage λcI protein. *Science* 263, 75–77.

42. Lin, L. and Little, J.W. (1989). Autodigestion and RecA-dependent cleavage of Ind-mutant LexA proteins. *J. Mol. Biol.* 210, 439–452.

43. Little, J.W. (1984). Autodigestion of lexA and phage λ repressors. *Proc. Natl. Acad. Sci. USA* 81, 1375–1379.

44. Little, J.W., Shepley, D.P., and Wert, D.W. (1999). Robustness of a gene regulatory circuit. *EMBO J.* 18, 4299–4307.

45. Liu, Y., Bonderenko, V., Ninfa, A., and Studitsky, V.M. (2001). DNA supercoiling allows enhancer action over a large distance. *Proc. Natl. Acad. Sci. USA* 98, 14883–14888.

46. Luo, Y., Pfuetzner, R.A., Mosimann, S., Paetzel, M., Frey, E.A., Cherney, M., Kim, B., Little, J.W., and Strynadka, N. (2001). Crystal structure of LexA: A conformational switch for regulation of self-cleavage. *Cell* 106, 585–594.

47. Nickels, B.E., Dove, S.L., Murakami, K.S., Darst, S.A, and Hochschild, A. (2002). Protein-protein and protein-DNA interactions of sigma70 region 4 involved in transcription activation by lambda cI. *J. Mol. Biol.* 324, 17–34.

48. Pray, T.R., Burz, D.S., and Ackers, G.K. (1998). Cooperative non-specific DNA binding by octamerizing λ cI repressors: A site-specific thermodynamic analysis. *J. Mol. Biol.* 282, 947–958.

49. Reinitz, J. and Vaisnys, J. R. (1990) Theoretical and experimental analysis of the phage lambda genetic switch implies missing levels of cooperativity. *J. Theor. Biol.*, 145, 295–318.

50. Révet, B., von Wilcken-Bergmann, B., Bessert, H., Barker, A., and Müller-Hill, B. (1999). Four dimers of a λ repressor bound to two suitably spaced pairs of λ operators form octamers and DNA loops overlarge distances. *Curr. Biol.* 9, 151–154.

51. Rusinova, E., Ross, J.B.A., Laue, T.M., Sowers, L.C., and Senear, D.F. (1997). Linkage between operator binding and dimer to octamer self-assembly of bacteriophage λ cI repressor. *Biochemistry* 36, 12994–13003.

52. Senear, D.F., Laue, T.M., Ross, J.B., Waxman, E., Eaton, S., and Rusinova, E. (1993). The primary self-assembly reaction of bacteriophage lambda cI repressor dimers is to octamer. *Biochemistry* 24, 6179–6189.

53. Shea, M.A. and Ackers, G.K. (1985) The OR control system of bacteriophage lambda—A physical model-chemical model for gene regulation. *J. Mol. Biol.* 181, 211–230.

54. Vologodskii, A.V., Levene, S.D., Klenin, K.V., Frank-Kamenetskii, M., and Cozzarelli, N.R. (1992). Conformational and thermodynamic properties of supercoiled DNA. *J. Mol. Biol.* 227, 1224–1243.

55. Whipple, F.W., Hou, E.F., and Hochschild, A. (1998). Amino acid-amino acid contacts at the cooperativity interface of the bacteriophage lambda and P22 repressors. *Genes Dev.* 12, 2791–2802.

Per Kraulis granted permission to use MolScript (Kraulis P.J. 1991. MolScript: A program to produce both detailed and schematic plots of protein structures. *J. Appl. Crystallogr.* 24, 946–950). Robert Esnouf gave permission to use BobScript (Esnouf R.M. 1997. *J. Mol. Graphics* 15, 132–134). In addition, Ethan Merrit gave us use of Raster3D (Merritt E.A. and Bacon D.J. 1997. Raster3D: Photorealistic Molecular Graphics. *Methods Enzymol.* 277, 505–524).

DESIGNING AN EFFICIENT DNA-BINDING PROTEIN

The following discussion is intended to illustrate by examples the factors that influence the efficiency with which a DNA-binding protein occupies its operator in the cell. For convenience the protein is a repressor, but the arguments apply generally. More precise and elaborate discussions can be found in the references cited at the end of the Appendix.

Synopsis

The fraction of time an operator is bound by a repressor is determined by two factors: the affinity of the repressor for the operator and the concentration of repressor free to interact with the operator. For any given affinity the occupancy is increased at higher concentrations of free repressor. The amount of free repressor in the cell can be significantly diminished by its interaction with nonoperator sites. One way to increase the efficiency of specific binding of a repressor is to replace a single operator site with two or more sites to which the repressor binds cooperatively.

Specific and Nonspecific Binding

We begin with a culture of bacterial cells, each of which contains an equal number of repressor molecules and each of which contains one operator. What fraction of the operators have a bound repressor at any given instant? We treat our cells as though they were fused into one large cell, maintaining the same concentration of operator and repressor as was found in the individual cells.

The dissociation constant K_{OP}, a measure of the repressor-operator interaction at equilibrium, is defined as follows:

$$OR \rightleftharpoons O + R \; ; \quad K_{OP} = \frac{(R)(O)}{(OR)} \tag{1}$$

where (R) and (O) = concentration of unbound (free) repressor and operator, respectively, and (OR) = concentration of repressor-operator complexes.

To determine the fraction of operators occupied by repressor for any given repressor concentration we derive an equation that gives us explicitly the fraction of operators free of repressor. If O_T = total operator, (O) = free operator and (OR) = bound operator, then $(O_T) = (O) + (OR)$, and substituting for (OR) in Equation (1) we can derive the relation

$$\frac{(O)}{(O_T)} = \frac{K_{OP}}{K_{OP} + (R)} \approx \frac{K_{OP}}{(R)} \text{ if } (R) >> K_{OP} \tag{2}$$

To see how this relation applies in a specific example, imagine a repressor present in 100 copies in an *E. coli* cell. Its concentration would be about 10^{-7} M, and the concentration of a single operator in the cell would be about 10^{-9} M. We assign the value of 10^{-10} M to the equilibrium constant K_{OP}.

The repressor is in large excess over the operator, and so, (R) ~ (R_T), the total repressor in the cell. Also, $(R_T) >> K_{OP}$, and so we can use the simplified form of Equation (2) to write

$$\frac{(O)}{(O_T)} \approx \frac{K_{OP}}{(R_T)} \approx \frac{10^{-10}}{10^{-7}} \approx 10^{-3} \tag{3}$$

This value for (O/O_T) means that, at any given instant, 99.9% of the operators in the population have a bound repressor. Or, put another way, each operator is bound 99.9% of the time.

The calculation of Equation (3) ignores the fact that proteins that bind tightly to specific operator sites also have a lower but not necessarily insignificant affinity for nonoperator DNA. This binding to nonspecific DNA sequences can significantly decrease the concentration of free repressor, and hence decrease the fraction of operators bound. In the following discussion, which illustrates this point, we treat repressor as having the same low affinity for all nonoperator DNA. It might be that nonspecific binding is dominated by sites that partially resemble operators, but this would not change our main conclusions.

The equilibrium constant, K_D, which describes the interaction of repressor with nonoperator DNA, is given by

$$DR \rightleftharpoons D + R; \qquad K_D = \frac{(D)(R)}{(DR)} \tag{4}$$

where (D) is the concentration of repressor-free nonoperator sites and (DR) the concentration of repressor-nonoperator complexes.

To determine the fraction of repressor bound to the nonoperator sites we note that if (R_T) = total repressor, then

$$(R_T) = (R) + (DR) + (OR) \approx (R) + (DR)$$

Substituting for (DR) in Equation (4) we derive

$$\frac{(R)}{(R_T)} = \frac{K_D}{K_D + (D)} \approx \frac{K_D}{(D)} \text{ if } (D) >> K_D \tag{5}$$

Imagine our repressor has a K_D of 10^{-4}M, not an unreasonable value. The concentration of nonspecific sites in a bacterium is about 10^{-2}M. (There are about 10^7 base pairs of DNA in the bacterium, and in this calculation we assume that each base pair represents the beginning of a new nonspecific site.) The number of nonspecific sites is always in vast excess over the number of repressors, and so we can set $(D) = (D_T)$, the total concentration of nonspecific sites. Then, from Equation (5) we find

$$\frac{(R)}{(R_T)} \approx \frac{10^{-4}}{10^{-2}} = 10^{-2} \tag{6}$$

The calculation of Equation (6) indicates that 99% of our repressor in the cell is not free in solution, but instead is bound to nonspecific sites and therefore is not free to bind operator. This in turn would mean that the fraction of operators bound by repressor was actually more than 100-fold lower than the value calculated in Equation (3).

In general, two ratios determine the efficiency with which repressor binds to the operator: K_D/K_{OP} and $(R_T)/(D_T)$. In our example, were the ratio of repressor's affinities for operator and nonoperator DNA to increase 100-fold (that is, were K_D/K_{OP} to increase 100-fold) the fraction of operators free of repressor would decrease about 100-fold. The same effect would be achieved by raising the repressor concentration 100-fold. To derive the general result, substitute (R) in Equation (2) with the expression for (R) given in Equation (5). Making a few simplifications we derive

$$\frac{(O)}{(O_T)} \approx \frac{K_{OP}(D_T)}{K_D(R_T)}$$

It is believed that, for at least some specific DNA-binding proteins, a significant fraction is sequestered on nonspecific DNA and is not free in solution in the bacterium. The problem is difficult to quantitate because of many uncertainties: for example, the affinities of a repressor for its operator and for nonspecific DNA measured in vitro may vary over several orders of magnitude as the salt and temperature are changed over reasonable levels, and we do not know, in any case, the precise ionic conditions that are relevant in vivo.

Increasing Specificity

Imagine now a would-be repressor with a ratio of specific to nonspecific DNA affinities such that, at a given concentration, it fills the operator with an efficiency of only 1%. How could we increase the efficiency to 99%? The following solutions might be imagined.

Increasing Protein Concentration

The concentration of the protein might be increased. As we have noted, about a 100-fold increase would be needed.

Improving Specificity Directly

The protein might be redesigned so that the ratio of specific to nonspecific binding would increase. We consider two ways to do this:

- The specific affinity is increased while the nonspecific affinity is held constant. We are not sure to what extent typical prokaryotic repressors could be improved upon in this way, but it would not be surprising if some effect could be obtained by adding an additional specific contact or two between one of the sequence-specific probes of the protein and DNA functional groups.

- The ratio of specific to nonspecific binding would be increased by increasing equally the number of specific and nonspecific contacts.

Consider a repressor that binds to its operator with a K_{OP} of 10^{-10}M and to random DNA with a K_D of 10^{-4}M. Now *double* the size of the repressor and of the operator, so twice as many specific and nonspecific contacts are made. The larger version, to a first approximation, would bind to operator and to random DNA with $K_{OP} = 10^{-20}$M and $K_D = 10^{-8}$M. [Twice as many contacts implies twice the energy $(-\Delta G)$ and recall that K is related to ΔG exponentially: $K = e^{-\Delta G/RT}$. For every 2.8 kcal change in ΔG, K changes 100-fold.]

Notice that by doubling the size of the repressor and of the operator we have increased the ratio of specific to nonspecific binding from 10^6 to 10^{12}. That is, the value K_D/K_{OP} has been increased by a factor of 10^6, and our analysis thus far would indicate that the larger repressor would work much more efficiently than does the smaller repressor. But now we run into a *kinetic* problem: the absolute value of the nonspecific binding constant is so high, and there are so many nonspecific sites, that the repressors spend virtually all their time on the nonspecific sites. That is, it would take a very long time to reach equilibrium between specific and nonspecific binding.

The kinetic problem arises because a repressor that binds to a site with a K_D of 10^{-8}M spends on the order of 0.1 second on that site. This estimate comes from measurements of the rates at which various repressors, once bound, come off their operators. We have noted that *E. coli* has in principle 10^7 nonspecific sites, and sampling even 10% of these would occupy our repressor molecule on the average over 25 hours, or more than ten times as long as a typical bacterial cell generation.

We surmise that our hypothetical doubled-size repressor would never find its operator during a cell generation. At equilibrium specific binding would be favored, but equilibrium would never be reached.

There is perhaps another problem raised by our double-sized repressor. A dissociation constant of $K_{OP} = 10^{-20}$M means that once a protein bound to the operator it would stay on virtually forever. No induction mechanism (nor any other cellular process) that required repressor to spontaneously fall off the operator would work within a meaningful period of time. Because the energy of binding is so large, the repressor would have to be drastically altered while on the operator to move it from the operator.

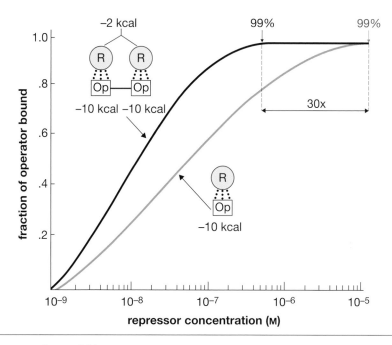

Figure A.1. It takes 30-fold more repressor to fill to the 99% level the single operator than to fill the double operator to which repressor binds cooperatively. Extensions of these curves would reveal that at the level of 99.9% binding, the difference in the required repressor concentrations would be about 100-fold. The curves were calculated assuming that the affinity of repressor for each site independently is –10 kcal, and the interaction energy between DNA-bound repressors in the two-site case is –2 kcal. In the two-site system, were the interaction energy –5 kcal, then, compared with the 2-kcal interaction case, the amount of repressor required to reach 99% and 99.9% would be 40-fold and 100-fold less, respectively. The two-site case in the figure approximates the filling of λ's O_R2 and O_R3 (in the absence of a functional O_R1), and the single-site case approximates the filling of O_R3 alone. (The two-site curve was calculated as the filling of one of the sites, but since the sites fill virtually together the curve is not changed much if we calculate instead the filling of both sites.) The curves were calculated by Sandy Johnson as described by Ackers et al. (1982).

Using Cooperativity

Cooperativity could be introduced into the system. For example, a second operator could be placed near the first one, and the repressor could be designed so that repressors bound to these sites would interact.

The curves of Figure A.1 show how cooperative binding to two sites would increase the efficiency of repressor binding, compared with binding to a single site. The curves assume that the repressor binds to a single site with an energy of interaction of –10 kcal, and that when bound to adjacent sites, the repressors interact with an energy of about –2 kcal. The curves show that about 30-fold less repressor is needed to reach the 99% occupancy level in the cooperative situation compared with the single-site noncooperative system. At the 99.9% level about 100 times more repressor would be required in the single-site case compared with the two-site system.

λ repressor, as explained in Chapter One, exploits the principles of cooperative binding. Were O_R1 the only site, or O_R2 the only site in the λ operator, the concentration of repressor in a lysogen is such that repressor dimers would fill the two sites to only about the 90% and 10% levels, respectively. But the measured –2 kcal of interaction energy between adjacent bound repressors raises the occupancy of O_R1 and O_R2 to over 99% in a lysogen.

Cooperativity, in effect, raises the ratio of specific to nonspecific binding; the probability of finding two repressors positioned on neighboring nonspecific sites is vanishingly small because of the large number of these nonspecific sites. In the λ case, two DNA-bound repressor dimers interact weakly and the protein-DNA complexes dissociate rather readily. In other cases there might be more interactions between DNA-bound proteins, perhaps involving several different proteins. Such interactions could produce highly stable complexes even though the individual proteins bind to DNA with affinities similar to those of λ repressor for its operator.

DNA-Protein Interaction in Eukaryotes

The nucleus of a eukaryotic cell is much larger than that of a bacterium, and it contains much more DNA. Therefore, a specific DNA-binding protein must be able to select its operator from among many more "nonspecific" sequences in the eukaryotic cell than in the bacterium. Does this present any special problems? We consider here one specific example that yields a simple answer.

Imagine a eukaryotic nucleus that bears one λ operator. Imagine also that this nucleus has both a volume and amount of DNA 500 times those of a bacterium. In other words, the total DNA concentration is the same in the eukaryotic nucleus and in the bacterium. It follows immediately that if the λ repressor concentration in this eukaryotic nucleus equals that in the bacterium—that is, if it contains 500-fold more repressor or about 10,000 molecules—then λ repressor would find its operator as well in the eukaryotic nucleus as it does in the bacterium. (Imagine a small volume of eukaryotic nucleus, equal in size to that of the bacterium, containing the operator and compare that directly to the bacterium. The ability of repressor to find its operator must be equivalent in the two cases.)

This simple description may approximate the actual situation in some cases. That is, the concentration of DNA in some eukaryotic nuclei is about that of a bacterium and, in the few cases tested, the affinity of eukaryotic regulatory proteins for their operators approximates that of λ repressor for its operator.

Our imagined scenario is a simplification designed to give the reader a starting point for thinking about the question posed at the beginning of this section. For any given case the concentration of the eukaryotic regulatory protein might be lower than we have imagined or the total DNA concentration higher. Either factor would make it more difficult for the regulatory protein to find its operator, a difficulty that might be overcome by cooperative binding.

FURTHER READING: RESEARCH ARTICLES

1. Ackers, G.K., Johnson, A.D., and Shea, M.A. (1982). Quantitative model for gene regulation by λ phage repressor. *Proc. Natl. Acad. Sci. USA* 79, 1129–1133.

2. Johnson, A.D., Poteete, A.R., Lauer, G., Sauer, R.T., Ackers, G.K., and Ptashne, M. (1981). λ repressor and cro-components of an efficient molecular switch. *Nature* 294, 217–223.

3. Lin, S. and Riggs, A.D. (1985). The general affinity of *lac* repressor for *E. coli* DNA: implications for gene regulation in prokaryotes and eukaryotes. *Cell* 4, 107–111.

4. Nelson, H.C.M. and Sauer, R.T. (1985). Lambda repressor mutations that increase the affinity and specificity of operator binding. *Cell* 42, 549–558.

5. von Hippel, P.H., Revzin, A., Gross, C.A., and Wang, A.C. (1974). Non-specific DNA-binding of genome regulating proteins as a biological control mechanism: I. the *lac* operon: Equilibrium aspects. *Proc. Natl. Acad. Sci. USA* 71, 4808–4812.

6. Yamamoto, K.R. and Alberts, B. (1975). The interaction of estradiol receptor protein with the genome: An argument for the existence of undetected specific sites. *Cell* 4, 301–310.

STRONG AND WEAK INTERACTIONS

The interactions of the components of λ's genetic switch are of two kinds: strong and weak. The protein-DNA interaction energies are in the range of –10 to –15 kcal (corresponding to a K_D of approximately 10^{-10} M) but the energies of the protein-protein interactions involve typically only –1 to –2 kcal (corresponding to a K_D of approximately 10^{-1} M). Thus the helping effect from protein-protein interactions amounts to about a tenfold increase in occupancy.

One implication of these facts is that, in vitro, effects that depend upon such weak interactions are easily overlooked. For example, using a simple transcription assay, P_{RM} is transcribed maximally at high polymerase concentration, and no stimulatory effect of repressor is observed. To take another example, if O_R1 and O_R3 are deleted, repressor at high concentration will bind to O_R2, and O_R1 might be deemed irrelevant. In each of these cases the effect of increased protein concentration can be mimicked in vitro by decreasing the salt concentration and thereby increasing the binding constant. Interpretation of these kinds of results is complicated by the fact that one does not know the exact conditions (concentration of interacting components, ionic strength, pH, etc.) that correspond to conditions in vivo.

If the roles of all the important components are to be revealed, therefore, it may be useful, as has been the case with the λ system, to challenge results obtained in vitro with experiments performed in vivo.

INDEX

Page numbers followed by an f indicate a figure.